The
JEFFERSON CITY
CIVIL PILOTS

★ ★ ★ *The* ★ ★ ★
JEFFERSON CITY
CIVIL PILOTS

From Lincoln University to Tuskegee Airmen

MICHELLE BROOKS

THE
History
PRESS

Published by The History Press
Charleston, SC
www.historypress.com

First published 2024

Manufactured in the United States

ISBN 9781467154499

Library of Congress Control Number: 2023945941

Notice: The information in this book is true and complete to the best of our knowledge. It is offered without guarantee on the part of the author or The History Press. The author and The History Press disclaim all liability in connection with the use of this book.

CONTENTS

FOREWORD

In 1866, a school on the northern edge of the Ozark Plateau was founded and funded by former slaves. More than 150 years later, that institution of learning still exists as Lincoln University of Missouri. This remarkable and unique beginning was followed by years of transformation, setbacks, resilience and progress. Many students, faculty, staff and administrators have passed through the doors of this "school on the hilltop," and the mission of its founders has remained the same: opportunity for education.

As a boy growing up in Jefferson City, Missouri, I lived in the neighborhood near Lincoln University of Missouri (LU) and spent three semesters at the LU laboratory summer school. This was my first true interaction with people of color. I had been raised in a household where respect for all people, no matter their origin, was taught and practiced, but going to the lab school opened my eyes to how the world really was—that people are just people. As I grew older, I continued my academic career at the local Catholic high school and then the University of Central Missouri, but I always looked back fondly on my experience at Lincoln. Several years later, I decided to become a schoolteacher and found myself back at Lincoln University, being offered an opportunity for education. After finishing my master's in history, I was offered a job as the night librarian at Inman E. Page Library on the Lincoln campus. Being in my mid-forties, I jumped at the chance of gainful employment, and it was the best decision I'd ever made. In 2009, I was offered the position I currently hold: university archivist.

The archivist's position at Lincoln includes facilitating education and outreach of the school's heritage. Before I took the position, I was familiar with the subject, but as I grew into the job, I began to appreciate the history of the institution and how it managed to not only survive but also flourish in later years. Part of the outreach portion of my position was to speak with members of the press. The local newspaper, the *News Tribune*, had a reporter who covered the Lincoln beat named Michelle Brooks. Ms. Brooks first approached me about an exhibit the archives was presenting in early 2010. My first impression of Michelle was that of someone who did their homework. After that encounter, Michelle and I kept in touch regularly, and she really helped to promote the school's image. She was inquisitive, professional and insistently looking for what lies underneath, a trait that has carried her far.

In January 2016, a student signed up for a library science independent studies class with the ominous title of "Theory and Practice of Research Methods and Data Management." I was to be the instructor, and the student's name was Michelle Brooks. This was a pleasant surprise! Michelle chose as her topic the 62nd Regiment of the United States Colored Troops, one of the two USCT regiments that founded Lincoln. I must admit that Michelle was quite impressive in displaying her research skills, and the assignment turned out to be the foundation for a future project for Michelle. Also, during that semester, she wrote a featured article about me for her paper. I am known on campus as a wordy sort, and she patiently listened to my extended monologues that really had nothing to do with the topic. What a kind soul! Of course, she passed with flying colors, and the work she did was outstanding. Michelle received her degree in Liberal Studies in 2018. During that time, she continued to work for the newspaper and investigate local history, compiling an impressive quantity of information and displaying a tremendous amount of savviness in locating facts, documents and images. I started learning from her.

Since then, Michelle Brooks has become a valued expert on local history in mid-Missouri. Her topics gravitate toward the hidden and little-known lore of the area, providing a refreshing slant on the heritage of the Jefferson City community. She has written a book called *Interesting Women of the Capital City*, featuring stories about known and unknown females from the local area. One of the things that Michelle has that impresses me the most is her presentation skills. She's presented on many topics, such as the 62nd Regiment, local African American cemeteries and the Tuskegee Airmen. Along with her spoken presentations, her graphics and images are not only professional but captivating as well. Michelle is someone who I am proud to know, and I have utmost respect for her skills.

In this project, Michelle Brooks addresses another long-lost subject, the Lincoln University Civil Air Pilots program. Since its founding, Lincoln University has had a deep connection with the U.S. military, including service in the Spanish-American War and World War I, the emergence of the on-campus military training program and, of course, the call to arms following the attack on Pearl Harbor in 1941. The sacrifices made by numerous young men and women to serve their country, one that considered them second-class citizens, are to be admired and celebrated.

Michelle's story is riveting, entertaining and informative. Painstaking research, a trademark of Ms. Brooks' past work, is evident here, along with fascinating facts and poignant stories. This book, as with her other works, addresses a topic that is an important part of American history but has not received the discussion it deserves. Michelle Brooks has resolved that.

Mark Schleer, Archivist,
Lincoln University of Missouri

PREFACE

In 2019, I came across a mention that Lincoln University of Missouri had the only Civil Pilot Training program for Black flyers west of the Mississippi River. I'd not heard of that, so I decided to look a little deeper.

I found out that in 1940, fewer than one hundred Black pilots had their license, mostly because they were denied opportunity at nearly all air fields at the time. Then I learned the value of the Civil Pilot Training program in preparing the woefully inadequate number of pilots in the United States for the inevitable World War II.

Next, I was introduced to the history of air fields in Jefferson City. The field the Jefferson City cadets trained on was the second one, but not the existing one.

Then, epiphany. Could pilots from Lincoln's program have become Tuskegee Airmen, the U.S. Army Air Corps' first opportunity for Black airmen? Immediately, a few names emerged. But I kept looking and discovered that dozens of Lincolnites were among the first Black men to serve in the U.S. Army Air Corps. This was a story I could not keep to myself.

The Jefferson City Civil Pilots: From Lincoln University to Tuskegee Airmen is a story of opportunity. Lincoln University of Missouri provided it, and nearly fifty men took advantage of it. Not all of the CPT cadets entered the U.S. Army Air Corps, but nearly all served in some military role. What is important is that these individuals were among the first few hundred licensed Black pilots in the nation. They were an inspiration to their race, although they may not have recognized their role in the Double Victory except in hindsight.

Lincoln had the vision to provide the CPT course. The CPT course offered evidence that Black pilots were just as capable as white pilots, encouraging the development of the 99th Fighter Squadron and the Tuskegee Army Air Field. Tuskegee built up pilots and ground crews who inspired a generation.

Several of these Lincolnites went on to break more racial barriers in their civilian lives after the war. This is a celebration of their lives, accomplishments and influence on racial progress.

ACKNOWLEDGEMENTS

First, I would like to thank the family members of these Lincoln flyers for sharing their memories and stories: Kathy Ayeh, Jacques Bordeaux, Richard Davis, Gina Fulbright-Powell, Adrienne Hoard, Kathleen Jingwi, Aziza Johnson, Percy Johnson, Luvanda Linebarger, Karen Long-Rowe, Naomi Long Madgett and Taalib-Din Ziyad.

This book, like the five before it, would not be possible without the input and encouragement from local friends, family and experts, including: Jeremy Amick, Dr. Christine Boston, Stephen Brooks, Ithaca Bryant, John Caples, Carl Haake, Cynthia Chapel, Lena Evers-Hillstrom, Janet Gallaher, Danisha Hogue, Gary Kremer, Charles D. Machon, Richard McGonegal, Amy Nickless, Tiffany Patterson, Bob Priddy, Terry Rackers, William Reagan, Nancy Arnold Thompson, Randy Turner, Janet Gallaher, Martin Warren Chapter of Missouri Society of the Sons of the American Revolution; and Frank Walleman.

As always, I am grateful to the keepers of documents, photos and stories so that future generations have access to the truth of our past, including: Nathaniel Arndts, Albion (Michigan) District Library; Jessica A. Black, St. Louis Lambert International Airport; Black Archives of Mid-America; Wil Curran, St. Louis University; Barbara Davis, New Rochelle Public Library; Jerome A. Ennels, Air Force Historical Research Agency; Kristina Hampton, St. Louis Science Center; Dr. Daniel Haulman, Air Force Historical Research Agency, retired; Adele Heagney, St. Louis Public Library; Kate Igoe, Smithsonian Nation Air and Space Museum; Grace

Lehner, Evanston (Illinois) History Center; J. Todd Moye, University of North Texas; David Pfeiffer; Cat Phan, University of Wisconsin–Madison; Michael Price, Springfield-Greene County Library District; Missouri State Museum; Patrick Raftery, Westchester (New York) Historical Society; Karen Raines, University of California–Riverside; Debora Richey, Fullerton (California) Heritage Board; Lauren Sallwasser, Missouri History Museum; Mark Schleer, Lincoln University of Missouri Archives; Becca Skau, Jackson (Michigan) District Library; Stephen Spence, National Archives at Kansas City; Patricia Smith Thurman; and Mark Wilderman, Scott Army Air Force Base (Illinois).

Finally, thank you to The History Press and Chad Rhoad for allowing this opportunity to tell Lincoln's unique story.

Chapter 1

THE FIRST TO FLY

Wilbur Long loved to fly. Growing up in the home of a teacher and preacher, education and character were priorities. But taking to the skies was his love.

That same fascination with the new field of aviation was shared by untold Black Americans in the 1930s. Discrimination kept most from going further than watching the skies. In 1939, only four African American pilots held commercial licenses, twenty-three held private licenses and eighty-two were students.

When Long heard that a Civil Pilot Training (CPT) program—the only one exclusively for Black flyers west of the Mississippi River—had been approved in Missouri, he was among sixty-five hopeful applicants for ten coveted slots. One year earlier, Black flyers had not even been considered when the CPT program was approved by Congress under the U.S. Department of Commerce and its new Civilian Aeronautical Authority (CAA).

Once accepted into the program at Lincoln University of Missouri in Jefferson City, Long transferred from Harris Teachers College, St. Louis, having taken flying lessons at the Curtiss-Wright School in East St. Louis, Illinois.

The first ten Lincoln CPT cadets, including Long, arrived at the old Jefferson Airfield (about where the Turkey Creek Golf Course is today) on September 16, 1940, for their first lesson. While they were young men infatuated with the new field of aviation, they also represented hope to an entire race.

That same day, September 16, 1940, President Franklin Roosevelt signed the Burke-Wadsworth Act, commonly known as the Selective Service Act. Among other changes, this act required each military branch to fill its ranks with minorities at the same percentage they were found in the general population. For these promising Lincolnites, and those who followed, that meant there would be a place in the U.S. Army Air Corps for Black airmen.

Wilbur Long was the first cadet to fly solo in the first class of Lincoln University of Missouri's Civil Pilot Training program. *Lincoln University Archives.*

In the decades since the Wright brothers made their historic sustained flight, myths abounded that African Americans had neither the aptitude nor the dexterity for flying machines. Just as they were excluded from so many other elite opportunities, the early aviation world tried its best to keep the door shut to Black flyers. Most stringent was the U.S. Army Air Corps, which had designed itself as entirely whites only.

The initial Civil Pilot Training Bill of 1939, signed by Roosevelt in June, was designed to build up the number of trained pilots who could then join the Army Air Corps, so planners initially ignored Black interests. Racial leaders from the Brotherhood of Sleeping Car Porters, the NAACP, the National Urban League and the Black press lobbied to ensure Black military aviation opportunities in the anticipated war in Europe.

The result was the Dirksen Amendment to the 1939 CPT bill, which allowed a few Black colleges as a test. North Carolina Agriculture and Technology, West Virginia State College, Hampton University, Tuskegee Institute, Delaware State College for Colored Students and Howard University offered the first CPT courses for Black pilots in September 1939.

By the end of the national CPT program's first year, the number of Black American pilots had doubled to 230. That success not only led to the additional CPT programs for Black pilots at Lincoln University of Missouri in 1940, but it was also an influential factor in the Burke-Wadsworth Act.

This monumental opening not only advanced racial equality in aviation, but it was also another rung on the ladder of Lincoln University of Missouri's pursuit to become the "Black Harvard of the Midwest." Already the institution had attracted high-caliber Black professors and was expanding its buildings, programs and student population. The elite CPT program fit right in.

The CPT program was part of the fledgling mechanical arts program, which had just moved into the new, state-of-the-art Damel Hall. G. Robert Cotton was in his third year, expanding the mechanical arts program with a variety of skilled trades.

Aviation was *the* new industry. The potential success for pilots, mechanics, engineers and other careers was obvious, and Cotton, along with university president Sherman Scruggs, was determined to see that Black students in the Midwest had a piece of the opportunity.

President Scruggs was one of the few Black army officers to emerge from World War I. He knew the slim possibilities for advancement in the military for Black men. It was his adamant focus that would bring a sought-after Black flight instructor to the Lincoln program.

However, before Lincoln could be considered for the CPT program, it needed an air field. The local air field, which had been created by the Jefferson City Chamber of Commerce a decade earlier, was managed by the Robertson Aircraft Corporation out of St. Louis. Conveniently, it already was host to the CPT program for Jefferson City Junior College cadets.

The second airfield in Jefferson City was located northwest of the present Jefferson City Municipal Airport, about where the Turkey Creek Golf Course is today. *Missouri State Archives.*

While most air fields across the nation excluded Black flyers, the local air field saw a chance for more government funding and expansion. The partnership between Lincoln University and the Jefferson Airfield was essential and mutually beneficial.

Future Tuskegee Airman Wilbur Long and his fellow cadets were the beneficiaries of these many moving parts falling into place for the CPT program to come to Lincoln University of Missouri. But it didn't stop there.

President Scruggs managed to lure one of only six licensed Black American flight instructors to teach this inaugural flight program for Lincoln. Recruiting C. Malcolm Ashe was "quite the coup for the Midwestern school" because the instructor was "one of the most successful and unusual pioneers of Negro aviation," according to the *New York Age*.

Ashe said that he chose Lincoln over other offers because "I was impressed with the magnitude and seriousness revealed in President Scruggs' personality." Even after courses began in Jefferson City, Black colleges in the East continued attempts to recruit Ashe. The national Black newspapers followed the Lincoln CPT program and Ashe attentively.

Four cadets from this inaugural CPT class became Tuskegee Airmen. These were the first generation of Black airmen in the U.S. Army Air Corps, so named because their segregated training facility was built near the Alabama town and school of the same name.

Almost fifty men completed Lincoln's CPT program; of those at least a dozen joined the Tuskegee Airmen. Even more Lincoln alumni also served with the elite Black flyers and their ground crew.

ON THE GROUND

Before Long and his fellow cadets could get in the air in the fall of 1940, they had seventy-two hours of ground courses, taught by Erskine Roberts, whose résumé was equally impressive as Ashe's. Roberts earned a Bachelor's of Mechanical Engineering degree from Northeastern University and a Master of Science degree from Massachusetts Institute of Technology.

Roberts first taught mechanical theory at Tuskegee Institute before moving to the faculty at the School of Engineering and Architecture at Howard University. He came to Lincoln in 1939, after serving as one of eight Black engineers appointed as inspectors/examiners on the Public Works Administration's nationwide construction.

Erskine Roberts was among the early faculty members of the Lincoln University Mechanical Arts program. He provided ground instruction for the Civil Pilot Training program. *Lincoln University Archives.*

Sadly, Roberts was injured just two months into the first CPT class when heavy machinery fell on him while being unloaded at Damel Hall, injuring his leg, lacerating his face and confining him to his home for more than a month. After Roberts' accident, the white flight operator and manager of the Jefferson Air Field, John Randolph, completed the remaining twenty-four hours of ground instruction for the Lincoln cadets.

Like Ashe, Roberts was in demand as the war effort heated up, and he became an expert in the production of B-29 Superfortress bearings. After the war, Roberts was a chief design and development engineer in the private sector before working for the Public Housing Administration; he ended his career as a top executive with a Chicago management firm.

IN THE AIR

Dressed in sharp military-style uniforms, the Lincoln CPT cadets received thirty-five hours of flight instruction from Ashe. "When some of the students were unable to get the feel of the controls, he had them remove their shoes and fly in stocking feet until they picked up the proper reactions," the *St. Louis Post-Dispatch* reported.

Their training was not without danger. For example, "One fellow had a scary experience when a control locked during a maneuver and the plane dropped to about 800 feet before, he regained full control," Ashe told the *Clarion*, the Lincoln University of Missouri newspaper.

It was national news in November 1940 when all ten cadets in Ashe's first class completed their solo flights. No surprise that the first solo flier in this inaugural CPT class was Wilbur Long. "It seems that everyone who gets interested in learning how to fly becomes exceptionally enthusiastic about it. Once the flying bug bites you, you are stuck with it forever. Wil was no exception," said Long's sister Naomi Madgett.

Long, who became one of thirty-two Tuskegee Airmen taken as prisoners of war (POW), was born in Virginia and spent his early years in New Jersey.

The inaugural class of Civil Pilot Training cadets at Lincoln University of Missouri included George Banks, Everett Bratcher, George Carter, Chester Cathorn, John Hughes, John Kincaide, Wilbur Long, James Merritt, Herman Plummer and Walter Sanderson. *Lincoln University Archives.*

He and his brother, Clarence, attended Harris Teacher College in St. Louis in 1939–40 after their father took up the pastoral post at a St. Louis Baptist church. Neither man wanted to become a teacher, but it was common at the time for Black St. Louis students to attend their first two years at the teacher's college to save money.

Wilbur Long left Lincoln without a degree the next semester to pursue an advanced CPT course at Chicago's Cornelius Coffey School of Aeronautics. Willa Brown and her husband, Cornelius Coffey, cofounded the first flight school owned and operated by African Americans. Brown also was the first African American officer in the U.S. Civil Air Patrol and was the first woman in the United States to hold both a commercial pilot's license and a mechanic's license.

By January 1941, nine of the first ten Lincoln CPT cadets had earned their private pilot's license, and plans were underway for the first all-Black fighter squadron in the U.S. Army Air Corps. The 99th Pursuit Squadron was the U.S. Army Air Corps' first step to comply with the Burke-Wadsworth Bill. Although a separate unit, it was still a new opportunity.

"Negro aviators will be accepted in the Army Air Corps....I consider my students good pilot material....If the army, which has a Jim Crow policy, refuses to enlist Negro pilots, the LU flyers hope to become flying instructors," Ashe told the *St. Louis Post-Dispatch.*

THE NEXT STEP

The men in the first Lincoln CPT class included future Tuskegee Airmen Everett Bratcher, James Merritt and John Kincaide. They were a reflection of the diverse heritage of those attending Lincoln at the time. They were often children of the Great Migration like these three, born in Arkansas, Tennessee and Mississippi, respectively.

John Kincaide was in the first Lincoln CPT class and was among the first Lincolnites to pass the U.S. Army Air Corps exams. *Lincoln University Archives.*

A building engineering major, Bratcher became the third Lincolnite to earn his pilot's wings in the U.S. Army Air Corps. Merritt, who earned a math degree from Lincoln, and Kincaide, an outstanding Lincoln athlete, both washed out of the military pilot program. Instead, they served as Tuskegee Airmen support crew, which averaged a 14:1 ratio of support to successful pilots.

At Lincoln, this pioneering class of CPT cadets continued their aviation enthusiasm the following semester by organizing the Storks Club, with the goal to "further individual knowledge of aviation and to contribute to the general welfare of the school." The club continued on campus at least through 1953.

The Storks Club's first president was Arkansas native Chester Cathorn. Interestingly, well before he was selected for the CPT program, Cathorn submitted to the federal government a diagram for a bombing device, which he called an "aerial torpedo." He was told that the government deemed it "undoubtedly of value."

Most of these first Lincoln CPT cadets were athletes and involved in campus fraternities or organizations, like Herman Plummer, a sprint star on the track team from Kansas. A few were nontraditional, like George Carter, who was married with three children and working full time at Warwick Village while attending the CPT courses. Nearly all of the CPT cadets who did not qualify for the U.S. Army Air Corps served in some military branch, like football star John Hughes, who joined the U.S. Navy, and biology major George Banks, who was deployed to the South Pacific with the U.S. Army.

From this first Lincoln CPT class, Walter Sanderson may have made the greatest impact during the war. The Fulton native became a professor at Lincoln, teaching National Defense Training classes to community members at Damel Hall.

C. MALCOLM ASHE

Approval for a second CPT course through Lincoln was delayed. The Jefferson City Junior College's CPT course had been running for a month before Lincoln was given the go-ahead in the spring of 1941. In the meantime, flight instructor C. Malcolm Ashe had moved on to other flight opportunities, leaving Lincoln's second CPT class to be taught by white air field manager John Randolph.

Dapper and affable, according to the *New York Age*, Ashe shared a paternal grandfather and his Cherokee heritage with tennis legend Arthur Ashe. Although he was born in North Carolina, C. Malcolm Ashe's family moved to Pennsylvania after his father died. There they all worked as domestic servants, but he did not let his circumstances deter him from a career in flight.

Ashe was the first African American to earn a pilot's license in the nation's capital, earning his flight hours at the Aviation Institute of the United States. The Howard University graduate added a commercial pilots license, a radio-telephone license, a flight instructor's license and memberships in the National Aeronautical Association and the Aircraft Owners' and Pilots' Association.

By October 1940, Ashe was nationally recognized. He tossed "a dead man's ashes to the four winds over a crowded city, [taught] classes of white students in an eastern school of aviation, [received] a degree in physiotherapy from the National College in Chicago, and [got] 30 offers from Negro and white schools to teach aviation within the last two months." He even provided a blood transfusion to a white pilot friend at a time when the U.S. military kept segregated blood banks based on white superstition.

From Lincoln, Ashe attempted to join the Royal Air Force Ferry Command. The RAF was looking for American pilots to fly out of Montreal, Canada, and would take pilots with as little as four hundred hours experience in any type of airplane. "I have about 2,400 hours and am a commercial pilot interested in receiving additional training," Ashe wrote. The RAF contracted with TWA airlines to provide the training at no cost to the pilots. Ashe said that his initial application to the TWA training center in Albuquerque, New Mexico, "received a favorable answer." Then his color was indicated when it was forwarded to Montreal. He did not hear from them for some time, so he personally visited the British Air Ministry in Washington, D.C.

That month, September 1941, the NAACP wrote a letter to Prime Minister Winston Churchill citing five types of discrimination in the RAF's bomber ferry pilot requirements. In March 1942, Churchill's office replied, "The British government has dropped the requirement that applicants as pilots to ferry bombing 'must be of the white race.'" The prime minister's office emphasized that this "was not in accordance with the views of His Majesty's government." But on Ashe's application in particular, the prime minister's office said that he was refused not because of his color but because "he did not state on his application that he had experience piloting multi-engine aircraft in darkness or in condition of poor visibility."

Eventually, Ashe landed as an instructor at Tuskegee. After the war, Ashe was a mechanic at Andrews Air Force Base, Maryland, where he "serviced presidential planes and investigated airplane accidents to determine their cause." Today, he is recognized on the Smithsonian Institute's Air and Space Museum's Wall of Honor.

Chapter 2

BEYOND THE SOLDIERS' DREAMS

The second Lincoln CPT class may have seen a delayed start, but two of its cadets became Lincoln's first commissioned pilots in the U.S. Army Air Corps. Wendell Pruitt and Richard Pullam were among fewer than five hundred Black pilots to fly combat in Europe.

Like Wilbur Long, Pruitt transferred from a St. Louis college to take advantage of Lincoln's CPT program. Kansas City's Pullam followed several siblings at Lincoln, including his brother, Arthur, a biology professor, and sisters enrolled in the home economics program.

The Lincoln University these young men arrived at was a place of promise for Missouri's Black citizens. Since the school was founded in 1866 by soldiers and officers of the 62nd and 65th U.S. Colored Troops, it had grown and advanced alongside Black Missourians, gaining skills, knowledge and respect.

Missouri had made it illegal to teach reading and writing to free or enslaved people of color in 1847. That was changed only after Missouri's 1865 constitution emancipated all Black Missourians and provided for their education.

While in military service from 1863 to 1866 in Texas and Louisiana, the founding Black soldiers learned to read and write from their white officers, who stressed literacy as an essential skill for good citizenship. The "Soldiers' Dream" was for the same opportunity to be available to anyone in Missouri, but especially to Black men and women.

Lincoln University of Missouri campus circa 1939. *Lincoln University Archives.*

The first courses, when the school opened in September 1866 as Lincoln Institute, were basic literacy, taught to both children and adults. Soon, the school took on the noble role of equipping Black teachers to return to rural schoolhouses across the state. In 1870, Lincoln began receiving state funds as a teacher training school for African Americans. By 1872, the school had built its first building on the Lincoln hilltop campus. Previously, classes had been held in an abandoned city schoolhouse and then in local Black church houses.

The school became a land grant institution in 1890, adding the now-prestigious agriculture department. As professional opportunities expanded for Black Missourians, Lincoln followed with more diverse courses, including an industrial school, teaching such skills as shoe repair, blacksmithing and carpentry.

Lincoln Institute became Lincoln University in 1921 when St. Louis dry cleaning business owner Walthall Moore, the first Black man to sit in the state General Assembly, introduced the bill elevating Lincoln to an equal status, albeit on paper only, as the whites-only University of Missouri–Columbia.

Accreditation as a four-year liberal arts school and pre-professional programs—including journalism, law and medicine—equipped Lincoln for its next generation of students seeking even more opportunity and advancement.

By 1939, Lincoln was filled with Black professors at the top of their various fields. Students attending Lincoln were the cream of the crop of their respective, though segregated, high schools across Missouri and from surrounding states.

President Sherman Scruggs

At this critical point in Lincoln's evolution into a modern, competitive institution, President Sherman Scruggs was at the helm. Born in Nashville, Tennessee, and reared in Kansas City, Scruggs earned a bachelor's degree from Washburn, taught agriculture at Tougaloo College, Mississippi, and became the first Black student to earn a doctorate degree from the University of Kansas.

Lincoln University of Missouri president Sherman Scruggs was among the few World War I Black officers and an advocate for the Civil Pilot Training program at the growing university. *Lincoln University Archives.*

Not only did Scruggs have the education and prestige for the role, but he also brought military leadership experience. Scruggs enlisted as a private in December 1917 and deployed as a corporal in April 1918 with the 370th Combat Infantry, the only combat regiment of World War I commanded entirely by Black officers. He distinguished himself as a sergeant major at the Battle of Belleau Wood in France in June 1918. While attached to a French infantry division, his "continued bravery" was noted in a French general's report, for "securing under heavy shell fire the liaison with contiguous units." For this bravery, he was awarded the Croix de Guerre medal.

A few months later, Scruggs was selected as one of 7 Black men from a pool of 2,900 to attend Army Candidates School in France. After being commissioned, he was assigned to the 365th Infantry Regiment and returned home a first lieutenant.

MECHANICAL ARTS AND PROFESSOR G. ROBERT COTTON

Professor G. Robert Cotton shared Scruggs' vision of a brighter future for Lincoln students through the mechanical arts program, which housed the Civil Pilot Training (CPT) program.

In 1938, G. Robert Cotton was hired as the only instructor in the

mechanical arts program. Within two years, the program had a new state-of-the-art building, Damel Hall, completed with state and Public Works Administration funds; a faculty from diverse backgrounds; and a degree program, which included practical and education degrees in building engineering and construction, mechanic arts, industrial arts and mechanical construction.

G. Robert Cotton was instrumental in expanding the new mechanical arts program at Lincoln University of Missouri, which included the Civil Pilot Training program. *Lincoln University Archives.*

The CPT program fit Cotton's view of Black education, "that the Negro's place in the economic pattern of American life will be determined largely by the Negro's ability to work with his hands," the *New York Age* reported. This viewpoint, notably promoted by Booker T. Washington, was one that Cotton widely was recognized for and carried from his previous work at Lincoln School,

The Lincoln University of Missouri mechanical arts program was housed in Damel Hall, which opened in 1938. *Lincoln University Archives.*

Atchison, Kansas. However, Cotton was aware that the prejudice within the trade unions and other matters related to construction and engineering were still a barrier.

By the spring of 1941, the mechanical arts program had experienced "phenomenal growth," from an enrollment of nine when he began to more than one hundred students. The *Age* called the program "one of the unsung, meritorious chapters in the development of Negro education."

In addition to developing four-year degree programs and supervising an ever-growing faculty, the university merged the campus building and grounds responsibilities into the mechanical arts program. Among the first projects was construction of the historic Mitchell Hall, housing the first journalism school in the nation taught entirely by Black faculty, as well as the English Colonial–style home economics cottage.

Cotton earned his Bachelor of Science degree from Hampton Institute in 1930 and a master's degree from Kansas State Teachers College, Pittsburg, Kansas. During his time at Lincoln, Cotton completed his PhD at Ohio State University with a dissertation on "Collegiate Technical Education for Negroes in Missouri with Proposed Plans for Development."

After the war, Cotton and the mechanical arts department built housing for nearly 180 veterans on campus. Cotton also negotiated with J.R. Brummet, the manager of the new municipal air field, in the fall of 1945 to continue a pilot training program.

From Lincoln, he took the president's post at the Kansas Technical Institute, Topeka, the only school of higher learning in Kansas employing Black faculty. Later, Cotton was appointed as education commissioner for the Virgin Islands, was a candidate for the head of Tuskegee Institute and served as interim president of Hampton Institute, Virginia.

SECOND LINCOLN CPT CLASS

The second CPT class at Lincoln was full of exceptional young Black men. Along with future combat pilots Pruitt and Pullam, Edwin Barrett also earned his wings at Tuskegee Army Air Field, although a medical complication prevented deployment overseas. And Winston Rogers, who came to Lincoln as an Olympic-level trackster, served at Tuskegee Army Air Field as a flight instructor.

Because of C. Malcolm Ashe's departure, their flight instruction came from the white air field manager John Randolph. Within three weeks, more than half the cadets had soloed. Upon landing, a cadet who completed his first solo flight was met "with a tub of water and [the] boys were lifted off their feet and headed into the water—a special treatment rendered to all LU soloists," the *Jefferson City News Tribune* reported.

The second Civil Pilot Training class at Lincoln included the "three musketeers," Bertran Wallace, Wendell Pruitt and Richard Pullam. Other members were James Barrett, Robert Buck, John Perry, Winston Rogers, Henry Scarlett, Compton Taylor and Julian Witherspoon. *Lincoln University Archives.*

In the midst of this class in March 1941, the cadets would have learned about the 99th Fighter Squadron ground crew training beginning at Chanute Field, Illinois. Among those first 250 Black airmen were two Lincoln alumni: Clovis Bordeaux and Robert White.

Regardless of where their futures went, members of this second Lincoln CPT class were trailblazers, being among the first one thousand Black Americans with the opportunity to fly. Other cadets included education majors Robert Buck and Compton Taylor, Chillicothe's John Perry, Georgia-born Henry Scarlett and St. Louis native Julian Witherspoon.

Perhaps the most qualified and capable member of the second Lincoln CPT class was Bertran Wallace, who came from Kansas City as its top ROTC cadet. He nearly received an appointment by Senator Harry Truman to a military academy. His charisma was apparent in his being elected class president all four years at Lincoln. And Wallace's athletic prowess resulted in multi-year, All-American nods for football and regional honors for basketball.

Wallace was convinced that military aviation was the best way to serve his country. When the call came for Black Civilian Pilot Training participants to join the U.S. Army Air Corps in the spring of 1941, he joined Pullam, Pruitt and other Lincolnites at Kansas City's Fairfax airport to complete the military entrance exams.

Wallace, Pullam and Pruitt called themselves the "three musketeers"—being friends, roommates and fraternity brothers. The first stop was the physical exam, which Pullam and Pruitt passed ahead of Wallace's turn. However, the same would not be true for Wallace.

"When the captain called me 'boy,' I indicated to him that my name wasn't 'boy.'...With that, he told me to get the—the use of the curse word F—out of here, and your F-ing days are over. And as a result of that, I cried for about five years. Because I felt this was where I could best serve my country," Wallace said in a Library of Congress interview.

While Pullam and Pruitt went on to fly, prejudice derailed Wallace from aviation. He spent the rest of his decades-long military career challenging racial barriers and shattering racial myths.

Chapter 3

A FIELD TO FLY FROM

The Civil Pilot Training program helped the nation raise its poor number of civilian aviators. With global conflict on the horizon, a pool of candidates with a familiarity with aviation would make a big difference. When the CPT bill was introduced in 1939, the U.S. Army had only 4,500 pilots, active and reserve.

The concept also fit with President Franklin Roosevelt's New Deal to "provide economic relief and encourage revitalization of industry.... Additionally, the CAA hoped these newly trained pilots would increase the demand for light aircraft and, in turn, stimulate the light-aircraft industry," as Major Renee Fontenot wrote in "Upward Mobility: The Civilian Pilot Training Program, War, and Society in the American Century."

By the end of its second year, the national CPT program had trained 9,000 pilots at 435 sites. As a result, the Army Air Corps' pilot numbers rose from 4,500 in 1939 to 27,000 in 1941. During its four-year existence, the CPT program operated at 1,132 colleges and universities, as well as 1,460 flight schools, including Alaska, Hawaii and Puerto Rico. "This all-out push for more pilots made it possible for hundreds of women and African-Americans to enter the somewhat exclusive world of flying," said Federal Aviation Administration historian Terry Kraus.

The Lincoln program shared training planes at the Jefferson Airfield with cadets from the Jefferson City Community College. *Lincoln University Archives.*

ABRAM JACKSON

Although the number of Black pilots was on the rise, the number of Black flight instructors was still minimal. Nevertheless, Lincoln president Scruggs and Professor Cotton hired a new Black flight instructor before the third Lincoln CPT class began flying over mid-Missouri in June 1941: World War I veteran Abram Jackson.

Jackson was a steward at the Erie, Pennsylvania, Aviation Country Club, where he earned his private pilot's license in 1937. Then he received his instructor's license in Chicago in 1939. That same year, Jackson was part of the first annual national air conference for Black fliers in Chicago, Illinois, where the National Airmen's Association of America was organized. He was elected assistant secretary under President Cornelius Coffee and Secretary Willa Brown.

Jackson also stayed only one semester, returning to Pennsylvania as an instructor at the New Alexander Airport in Pittsburgh before being assigned as a ground instructor at Tuskegee Army Air Field.

THIRD CLASS

While the third Lincoln CPT cadet class was underway, the first class of aviation cadets for the U.S. Army Air Corps' new 99th Pursuit Squadron began training at Kennedy Field, near the future Tuskegee Army Air Force base, which had yet to be constructed.

Milton Lemmons was this class' only Tuskegee Airman.

Cadet Ernest Bennett took his pilot training with him when he returned to his hometown of Oklahoma City, Oklahoma, to teach math. One year later, Douglass High School added an aeronautics course taught by Bennett, who had earned his instructors rating. Two other Oklahomans took the CPT class that summer: Lance Barber and James Garrett.

Other trailblazing cadets from Lincoln's third CPT class included Frank Bruce and Alphonse Ellis, who would each serve as president of the Alpha Phi Alpha chapter on campus during the war; *Clarion* writer Henry Savage; graphic arts major Earl Hogan, who later served in the U.S. Navy; George "Rabbit" Thompson, who served in the 743rd Military Police (Aviation Battalion); and Richmond native Glenwood Johnson, who continued his service as an officer in the Missouri National Guard after the war.

Cadets in the Lincoln University of Missouri Civil Pilot Training program during the summer of 1941 included Lance Barber, Ernest Bennett, Frank Bruce, Alphonse Ellis, George Thompson, James Garrett and Henry Savage. *Lincoln University Archives.*

JEFFERSON AIR FIELD

As early as 1917, young Black men were attempting to enlist in the air service. Excuses from the war department included a lack of Black officers and the myth that Black men were not attracted to aviation in the same way as white men. The CPT program went a long way toward disproving that myth and opening the door to the U.S. Army Air Corps.

A key to Black Missourians breaking through the military's aviation racial barriers was Jefferson Airfield. Most private air fields and flight training organizations followed the similar false logic as the military, that African Americans were somehow physiologically and psychologically ill-equipped for aviation.

Local airport manager John Randolph did not agree. Although the Jefferson City Junior College and Lincoln programs were separate courses, the students could be found sharing class time and instructors. "No other nation in the world has ever enjoyed such a large reservoir of private fliers

The Civil Pilot Training program at Lincoln University of Missouri was possible only through the cooperation of the Jefferson Airfield, operated by Robertson Aircraft Corporation out of St. Louis. White students from the Jefferson City Junior College were taught by white flight instructors. But some ground classes were integrated. *Missouri State Archives.*

and private airplanes available for military use and training....After all, the first 50 hours a man spends in the air is spent merely at learning how to look at the ground. These little ships do that efficiently, safely and many times cheaper than larger bulkier training planes," Randolph told the *Jefferson City Post Tribune*.

Light planes with a minimum fifty horsepower, though not military type, were used for primary flight instruction at the Jefferson Airfield. The textbook provided by the Civil Aeronautics Authority, *The Primary Ground Study Manual*, included topics such as the history of aviation, theory flight and aircraft, parachutes, aircraft power plants, aircraft instruments and air traffic control procedures and phraseologies.

The curriculum came in three stages. First the student received a minimum of eight hours of instruction in aircraft familiarization, taxiing, air maneuvers, takeoffs, landings, spin recovery and simulated forced landings. Next, he conducted a minimum of three hours of solo work. Finally, the cadet performed a minimum of fifteen hours of solo work and eight hours of dual work with an instructor, including precision landings, advanced stalls, cross-wind takeoffs and landings, power approaches and landings and completed cross-country flight.

The national CPT program added an advanced flying course with more ground instruction in navigation, radio operations and cross-country and night flying, as well as a seaplane program on the coasts. But the local airport's proximity to the northern limestone cliffs made it ineligible for the night flying instruction.

The air field was the result of the Jefferson City Chamber of Commerce seeing the need for such a facility and leasing eighty acres from Paul Koch in Callaway County. At the time, it was called "one of the finest natural airports that it is possible to have." It was "readily found by flyers who can identify it through its close proximity to the river and through the 100-foot-wide, stone circle which marks it," the *Jefferson City Post-Tribune* reported on August 3, 1929. William Robertson from St. Louis won the bid to manage it in 1938.

Newlyweds John Randolph and his wife, Martha, arrived in December 1938 as airport manager and secretary, respectively. Randolph previously flew for Robertson Aircraft Corporation out of Lambert Field, St. Louis. The Jefferson City site had one private hangar when they arrived. Within months, they had built a school with support facilities, including hangars, fueling, lodging and meals, taxi service to the city, repair facilities for minor emergencies, wind indicators, fire apparatus and first aid, as well as lighting

When Robertson Aircraft Corporation took over operation of the Jefferson Airfield, it immediately made improvements, including amenities for pilots and a flight school, all under the charge of manager John Randolph. *Missouri State Archives.*

with a beacon, boundary approach and flood lights. The Flying Boots restaurant was supervised by Mrs. Cecil Bowman.

"It was largely through the efforts of Manager Randolph that Jefferson City was put on the aviation map. When he came to Jefferson City, the airport had proved a poor investment and was almost on the rocks," the *Post-Tribune* said.

The CPT program presented another way for Randolph to improve the airport's prestige. Students praised him as "always helpful and courteous to them in their efforts to win their wings."

Randolph was a graduate of the Kemper Military Academy, Boonville, and attended Washington University, St. Louis. He flew his first solo flight in 1927. Before famed pilot Charles Lindbergh flew across the Atlantic, Randolph was a line service man working toward his wings at Robertson Aircraft Corporation, St. Louis, where Lindbergh was just another pilot for the firm.

The CPT program was "of inestimable value in building up this nation's defense," Randolph said, as the students were months ahead of raw recruits, and the cost of eliminating those unfit for flight duty was therefore reduced. The CPT program offered "us a head start toward manning the 50,000 airplanes the president wants," he said.

Chapter 4

THE FINAL CPT CLASS

When the fall 1941 semester began, no one knew that the May 1942 yearbook, *The Archive*, would be titled "The Year of Crisis."

The new school year likely was full of excitement and anticipation, particularly for the men selected for the fourth CPT class. They had seen earlier CPT cadets pass the U.S. Army Air Corps exams and knew of Lincolnites already serving in the 99th Fighter Squadron.

While these Lincolnites were taking off from one of three runways at the Jefferson Airfield, the first class of basic training cadets at Tuskegee Army Air Field was using one unpaved runway with construction all around. Three cadets from this CPT class became Tuskegee Airmen. Louis Harris, Asbury Gary and Robert Taylor all served in ground support roles during the war after washing out of the pilot training program at Tuskegee.

One month into the CPT course, Tennessee-born James Butler left the program, and Lincoln, to enlist. After the war, he returned to Lincoln, like hundreds, to complete his degree. Butler completed elementary flight training in the new pilot program offered through Lincoln at the new local municipal airport.

Others who completed the fourth Lincoln CPT course were journalism major William Graham, New Madrid native Lucius Banks and future chemistry teacher George Outlaw.

The fourth Civil Pilot Training class at Lincoln University of Missouri was held in the fall of 1941. Cadets included Lucius Banks, Asbury Gary, William Graham, Louis Harris, George Outlaw, Robert Taylor and James Butler. *Lincoln University Archives.*

FLIGHT INSTRUCTOR RICHARD DAVIS

This fourth CPT class benefited from flight instructor Richard Davis, who was hired by the Robertson Aircraft Corporation, St. Louis, after graduating from Hampton Institute, Virginia, where he completed the CPT course.

Born on a Georgia farm, Davis had an early curiosity for the crop dusters flying overhead and learned at a young age how to build and maintain an airplane. According to his nephew and namesake, Davis was an accomplished pilot by age sixteen.

Davis stayed at Lincoln only one semester, heeding the call for pilots at Tuskegee after the attack on Pearl Harbor. Earning his wings at Tuskegee with class 42G in August 1942, Lieutenant Davis was one of the original twenty-nine pilots of the 99th Fighter Squadron. The "cream of the crop," these were intelligent, physically fit young men from eighteen states.

The original 99th pilots were a tight-knit bunch. They "believed none of us is ready until all of us were ready," Lieutenant Davis' nephew said.

Lieutenant Davis struggled with some of the academics of the flight training and received help from his cohorts in the barracks basement after hours, so that they all graduated together. By graduation, Davis had reached the top of the class. "They laid the foundation for all that came after them," Davis' nephew said on the podcast *We Got a Story to Tell.*

Sadly, Davis was killed in a flying accident at Tuskegee on January 30, 1943, while the 99th waited to be deployed.

DAY OF INFAMY

The news of the attack on Pearl Harbor and President Franklin Roosevelt's "Day of Infamy" speech reached many former CPT cadets, who were traveling with the Lincoln football team aboard a train returning from their defeat at Wilberforce University, according to Jack Gibson.

"On the train back after the game, we heard the news from the Pullman Porters about the 'infamy' broadcast. That changed life for all of us. We all tried to enlist....I thought [being a pilot] sounded glamorous. I decided I'd look clean in a 'flyboy' costume with a silk scarf around my neck and the goggles and pants that stuck out on the sides, just like the aces I'd seen in the movies. I could just see myself landing my plane and gathering up all the pretty French girls."

Gibson, who was the grandson of Anderson Schweich, longtime food delivery and service provider and namesake of Schweich Hall on campus, was sidelined in his enlistment by poor eyesight. Several other students and alumni immediately responded to the declaration of war, including John Hughes from Lincoln's first CPT class, who joined the U.S. Navy on December 10, 1941.

SPRING OF 1942

Before the CPT program transformed into a military training program, Lincoln offered its fifth and final CPT class. By January 1942, the U.S. Army Air Corps added the 100th Fighter Squadron, tripling the pilot class sizes at Tuskegee from ten to twenty-nine. This meant that more Black men would have the chance to earn their wings.

Four cadets from this final class moved on to the U.S. Army Air Corps. Lewyn Boler, Lester Davis and Chicago native Wilbur George served in

The last Civil Pilot Training class at Lincoln University of Missouri included Lewyn Boler, Curtis Branch, William Crump, Joe Gayles, Wilbur George, Elmore Nelson, John Pope, William Spencer, Khamalaw White and Lester Davis. *Lincoln University Archives.*

support roles. Curtis Branch served at the U.S. Army Air Force base in Frederick, Oklahoma, with the B-26 "Marauder" medium bombers.

The remainder of this final class included Elmore Nelson, who served in the 743rd Military Police (Aviation) Battalion; Oklahoman William Crump; swing saxophonist Joseph Gayles; John Pope; and broad-jump champion Khamalaw White.

Also in this class was Cadet William Spencer Jr., grandson of early Lincoln supporter and soldier in the 68th U.S. Colored Troops James Spencer. His grandfather was the first Black doctor in St. Louis and is buried at Woodland-Old City Cemetery. Cadet Spencer's father graduated from Lincoln High School and then Lincoln Institute in 1899. His sister, Sarah, was a history professor at Lincoln for more than twenty-five years. Cadet Spencer received a music degree from Lincoln, being proficient with at least seven instruments.

About "50 young men have had this experience in aviation and have either a private aviation license or are working toward a license. The experience that these trainees are receiving is proving valuable for the providing of a reservoir of partially-trained persons for the armed air forces and for the promotion of commercial and private aviation," the 1942 yearbook said.

First to Tuskegee Army Air Field

Four months after the attack on Pearl Harbor, the call came from Tuskegee Army Air Field for the first of Lincoln's CPT cadets—Richard Pullam, Wendell Pruitt and Asbury Garry.

They were met there by two Lincoln alumni serving in the 99th Pursuit Squadron's ground crew, Clovis Bordeaux and Robert White. Both arrived at Tuskegee Army Air Field just before the Pearl Harbor attacks, as part of the first enlisted men of the 99th Fighter Squadron. St. Louisan Bordeaux hated flying and heights but chose the military for its technical opportunities and its steady paycheck. Born in Arkansas and reared in Chicago, Illinois, White made the military a career, reenlisting for another fifteen years.

By June 1942, more than forty Lincoln students were in service, spread out across sixteen forts and camps. The *Archives* yearbook was so dedicated: "To those who have gone forth to fight, forgetting private fate; to Lincoln's quiet sons, taking their places to beat back the night. Beneath the deadly fury of the guns, let each page be, a consecrated candle to keep bright, their names who in this desperate day unit, that all men shall be free."

Double Victory

The war effort at Lincoln and through its alumni was more than just retaliation for Japanese invasion or in support of allies threatened by a madman. The greater goal was Double Victory. As the yearbook said:

Lincoln is a small, but vital center in a nation aroused at last to an awareness of the exacting demands of Democracy's life-struggle; awaking to the truth that freedom cannot be successfully pursued as a basic war aim abroad, yet denied as the fundamental right of every citizen in our own country.

Lincoln men have been absorbed into America's armed forces and are helping industry to build Democracy's greatest arsenal; Lincoln women have been called to the nerve centers of our nation to lend their skill to the accomplishment of the job which lies before us.

Within the confines of segregation, nearly everyone on the Lincoln campus found a way to serve. The Young Women's Service Organization had formed before the attack on Pearl Harbor under the leadership of Mrs. Sherman Scruggs. The female students provided companionship and entertainment for the Black soldiers training at Fort Leonard Wood. They provided singers, drama and speakers. They traveled by bus in large, chaperoned groups to the post for dinners, dances, games and movies.

On campus, a Defense Council was also organized. One of its early objectives was providing courses that would "directly contribute to the demand for skilled workers in the war effort," as history professor Antonio Holland noted in the 1991 *Soldiers' Dream Continued: A Pictorial History of Lincoln University of Missouri*. Courses included foods, first aid, machine shop and secretarial work. Later, the council organized bond fundraisers, collection drives and other educational projects.

Chapter 5

WAITING FOR THE CALL

The fall of 1942 must have been different compared to the start of the previous school year, when war had not reached the steps of Young Hall. Students and faculty continued their work of education, but with the war in the back of their minds. Most of the men were eligible for the draft, if not eager to find their place voluntarily among the patriot ranks.

Many Lincoln students, hoping for the Army Air Corps' call, received deferments on their enlistment to continue their studies until openings became available. The addition of the 332nd Fighter Group—including the 100th, 301st and 302nd Fighter Squadrons—and plans for the 477th Bombardment Group further expanded opportunities for Black men to serve as airmen.

At that time, only 2.8 percent of the U.S. Army Air Corps personnel were Black, compared with the general population, which was about 10 percent. They had a long way to go to meet the standards of the Burke-Wadsworth Act.

For several pilot hopefuls, they aged out or were called up by their local draft boards before receiving their letter from Tuskegee.

FIGHTING FOR DOUBLE VICTORY ON CAMPUS

While the flyers and military members faced ongoing barriers based on racial myths within the military command structure, the students and faculty at Lincoln were standing up to discrimination in Jefferson City. This dual effort during the war years was striving for "Double Victory."

The thriving Black commercial district at "The Foot" of Lincoln's hilltop campus provided places to eat, play pool, listen to music, see a barber, grab a taxi, seek medical attention, buy groceries, have shoes repaired or get clothes dry cleaned. For nearly twenty years, The Foot had been growing to serve primarily local Black families, the majority of whom had homes within walking distance.

Several blocks west of the campus was the South Side neighborhood, established by primarily Protestant, German-speaking immigrants in the last half of the nineteenth century. This commercial district, along Dunklin and Jefferson Streets, also called Old Munichburg, mostly welcomed African Americans to shop in their stores or work in their community.

Northwest of the campus, however, was the uptown area, with origins dating back to 1823, when the primary residents and business owners were settlers from southern states. The stores and restaurants surrounding the Missouri capitol served Black and white customers differently, including the movie theaters, where limited, separate seating served both Lincoln students and all of the Black community. Most restaurants would take their money for a meal, but not let them eat inside.

UNWANTED PARTICIPANTS

In October 1942, the Jefferson City American Legion Post No. 5 organized a parade to raise money for war bonds. The university's band and civil defense classes, as well as the Toney Jenkins American Legion post and businesses from The Foot, were invited to participate.

Given the common mission, Black leaders hoped that they would be treated equally. The civil defense class expected to line up with similar community defense preparations. The Legion post expected to march with its white counterparts. Instead, all Black entries were relegated to the back of the parade. So, President Scruggs pulled all school participation. In solidarity, the other African American community entries followed Scruggs' move, according to the October 23, 1942 *Clarion*:

> [Scruggs] *issued the following statement: The ready and willing enlistments of our young men into the armed forces of the nation, the serious study of our young women to prepare themselves for the many areas of service on*

the home fronts, and the willing sacrifice that is often made to purchase bonds and stamps by older adults and children are greater manifestations of our loyalties and patriotism as really true Americans and good citizens of Jefferson City than marching as unwanted participants in a parade.

President Scruggs received letters and telegrams from across the state and nation in support of his withdrawal on principle. Attorney Carl R. Johnson, president of the Kansas City NAACP, wrote, "At long last Lincoln's leadership is definitely sound. Your action should be sound inspiration and guidance to all well-thinking persons in our state."

At the same time, President Scruggs instructed the student body to make "no violent moves or conspicuous outbursts regarding the wrongs that certain Jefferson City officials have heaped upon us....[explaining] our bearing and deportment today can do more to help or to defeat us than any other weapon."

The Lincoln campus and the surrounding African American neighborhood held its own war bond campaign and parade as part of "Army Day," the weekend before Armistice Day, November 11, 1942. President Scruggs put the planning in the hands of English professor and *Clarion* advisor Hazel Teabeau, who later was one of the first nine Black students to attend the University of Missouri–Columbia and became the first African American to earn a PhD there.

A "mammoth" parade with the theme "Education, the First Line of Defense" proceeded from the campus to the capitol steps. The band in uniform, a small corps of U.S. Army soldiers and the campus queen led the parade. Four banners followed President Scruggs, reading "Freedom of Speech," "Freedom of Religion," "Freedom from Want" and "Freedom from Fear." After that, nearly every organization on campus participated, including the Red Cross, safety education, air raid wardens, home management, public health, defense classes, PE, police and fire. The Greek organizations were in charge of the "Double V" project, which included a large double V mounted to a truck, reflecting the hopes of victory abroad and victory at home over racism.

At the capitol, students sang the national anthem and the school song. Following the parade's return, the Tiger football team faced off against the army team from Fort Sill, Oklahoma, whose quarterback, Willard Johnson, was a former Lincoln athlete. CPT cadet Bertran Wallace, who was inducted into the Lincoln Athletic Hall of Fame in 2011, was Lincoln's team captain. He was "one of the best ends in conference history. He stays on the field

Most of the future Tuskegee Airmen from Lincoln University of Missouri were athletes, playing on the football or basketball teams or running track. *Lincoln University Archives.*

almost 60 minutes every game. He will be a constant obstacle," the *Clarion* said. Other CPT cadets, like halfbacks George Thompson and Khamalaw White, distinguished themselves on the football field as well, while waiting their turn in military service.

At halftime, the names of 118 Lincolnites in the military service were read, and "Taps" was sounded for Richard Parker, the campus' first alumni known to be killed in action, dying aboard the USS *Lexington* in May 1942.

That evening, the Lafayette Service Men's Club presented "Cavalcade of Rhythm" at the State Park (McClung Park today). A dance followed, with music by the ten-piece Fort Leonard Wood Quartermaster Band. The bond drive was a success.

INVITATION WITHDRAWN

Only one month later, another wave of prejudice hit the campus in November 1942 when the invitation to join a national honorary scholastic fraternity was revoked.

Professor James Parks, head of the art department, received an unsolicited invitation from the Kappa Pi art society to set up a local chapter. At the time, Lincoln and Howard University were the only two historically black colleges to belong to the College Art Association of America, and the Lincoln program had eighty-five students enrolled.

Parks immediately began preparations for an honorary art fraternity chapter at Lincoln. One week later, Marie Bristow Ryan, Kappa Pi expansion officer, sent another letter from Kearney, Nebraska, saying, "When my secretary wrote the first contact letter to you…we did not have a catalogue of your university at hand, nor had we studied it very carefully, when we sent the second letter.…Today, I went over the catalogue very carefully and made the discovery that the institution was designed for the special benefit of the colored race. I regret to say that our constitution has specified that we establish chapters on college campuses designed especially for the white race. It is pleasant to know of your interest in an art fraternity and I hope that you will be able to find another one that is satisfactory."

In reply, Parks wrote:

> *It seems that Kappa Pi is much in need of expansion in broadness, democracy, tolerance and real old Christianity, rather than in expansion of chapters. Happily, artistic genius does not recognize artificial barriers of race and color.…We are fighting a great war for freedom in opposition to fascism and race oppression as epitomized by Hitler, while at home we find fascism, race intolerance and oppression in our own institutions.…Half a million Negro youth are fighting, dying and shedding their blood for an American democracy which, in most instances, is restricted to the "white" race just as Kappa Pi is.*

RACIAL COOPERATION

The City of Jefferson had much room to improve on in its race relations. A light of hope came in support of building the Jefferson City Community Center at 608 Dunklin Street. The Jefferson City Community Center Board had been meeting needs within the Black neighborhood through a food pantry, playground, a nursery school and other social services for more than a decade. It needed a permanent structure—today, one of two remaining buildings of the Historic Foot District. Construction was highly limited during the war years, yet this project was approved and received the unique blessing from the Community Chest (forerunner to today's United Way of Central Missouri) to make an appeal to the community-at-large for help to cover the building loan. The center was a beneficiary of the annual chest campaigns but had never made an appeal to the general public before.

Lincoln mechanical arts student Rolland Cooper designed the building with the instruction of Lincoln architect L.E. Fry. The son of a World War I veteran, Cooper had been an Eagle Scout in Springfield and helped Hubert

Washington, the university barber, organize the Jefferson City Negro Boy Scouts of America Troop No. 15 in early 1942. The twenty-one members of the troop, sponsored by the American Legion Toney Jenkins Post, met at Damel Hall. Cooper later served as a corporal in the U.S. Army and then worked as an engineer in California.

At the building dedication in November 1942, President Scruggs delivered an eloquent speech. Before that, fellow World War I veteran and Lincoln vice-president W.B. Jason presented an American flag to the center on behalf of the Toney Jenkins Post American Legion. Scruggs declared that three victories had been achieved through the opening of this center: "racial cooperation, community relations and acquisition of a facility designed to inspire."

The Jefferson City Community Center served as a USO for Black military men, as well as provided community services like organizational meeting space, a dance hall, soup kitchen and a place for cultural experiences.

Although work remained to create healthy race relations in the Capital City, the grateful community center board printed the following ad in the *Daily Capital News*: "America is the only place in the world where a minority can ask a majority for aid and get it."

WAITING FOR THE CALL

The campus saw at least two notable visitors in the fall of 1942. One was Harlem Renaissance poet and playwright Langston Hughes. Incidentally, this Joplin native worked as a busboy at a Paris restaurant owned by Eugene Jacques Bullard, the first African American military pilot to fly in combat during World War I.

Perhaps of more interest to the young men on campus at the time was Lieutenant M. Unterman of the Victorville Air Base, California. Unterman addressed two hundred men, providing firsthand information on military methods of operation. Nineteen Lincoln students enlisted that very day. Qualifying for the air corps enlisted reserve were CPT alumni Frank Bruce, Louis Harris, Wilbur George and Clinton Thompson. "Men enlisted in the air corps for active duty, who have not received their calls, were advised to remain calm, remembering that new fliers can be called only when vacancies occur," the *Clarion* reported.

While the eager airmen hopefuls had to wait, the draft was extended in November 1942 to include eighteen- and nineteen-year-olds. At the time, 124 Lincolnites had entered service. Campus speculation proposed that the school could become nearly all girls by the spring semester, with only 10 male students aged seventeen. The campus had 78 male students, aged eighteen and nineteen, and 138 were ages twenty to thirty-seven.

MAIL CALL

The *Clarion* campus newspaper was a mainstay for the service members each week during their mail call. They returned many letters to the editor from across the world to stay in touch during the war years. "News from the home front is what keeps us soldiers free from worries and loneliness," Lincolnite Clifford "Jewel" Shannon wrote from the 48th aviation squadron in Salt Lake City, Utah, in December 1942.

The *Clarion* was established in 1932 by the Kappa Alpha Psi fraternity as a bimonthly, eight-page product to stimulate intelligent student opinion and establish high standards of conduct, scholarship and sportsmanship. The paper received an upgrade when it became part of the new School of Journalism, which opened on campus on February 5, 1942, under Thelma Berlack Boozer, a nationally recognized newspaper executive. Mitchell Hall on Dunklin Street was built by the Mechanical Arts program specifically to house the soon-to-be prestigious journalism program.

The Lincoln School of Journalism was the "first all-Negro school for journalism" in the country. The journalism school was among several changes at the growing Lincoln University and the result of the lawsuit by veteran Kansas City journalist Lucille Bluford to gain admittance into the University of Missouri–Columbia's journalism graduate program.

The *Clarion* was a voice for the injustice that students and the community were experiencing. For example, it tackled the long-standing issue of limited and segregated seating at the white-owned movie theaters uptown. Black students were growing more frustrated with their limited access to the two local movie theaters. One did not allow Black seating at all, and the other limited seating to seventy-nine seats in a small balcony.

A well-written editorial in the December 4, 1942 *Clarion* encouraged students to find an alternative to returning "each week asking for more humiliation" at the town theaters. They proposed that a second movie could be shown on campus on Wednesdays, as well as the Sunday show, at Page Auditorium. "To remedy this situation is our problem. To solve this problem is to take the sting from the claws of Jim Crow."

Many future U.S. Army Air Corps members worked on the *Clarion* staff, including Wendell Pruitt as sports columnist, Louis Harris as editor-in-chief and Bertran Wallace as technical adviser. The staff photographer during the war was Clarence Long, brother of Tuskegee Airman Wilbur Long, who became the first graduate of the journalism program.

DEFENSE COUNCIL

Like most schools, Lincoln organized a Defense Council, which included pre-induction training for military service, scrap collection and war bonds/stamps.

One of the great losses in the salvage effort was the iconic cannons that had stood in front of Memorial Hall. "The cannons' value to the appearance of our campus cannot be as significant as their value to our war production. Our cannons are now on the scrap pile at Chestnut and Atchison streets. They are now a part of Lincoln's contributions to the war effort. Hooray!" noted a letter to the editor in the October 1942 *Clarion*.

To address fuel rationing, the basketball team—which included several future Tuskegee Airmen—reduced travel to weekend tournaments. And to address food rationing, meatless Tuesdays began well in advance of the federal compulsory requirement in December 1942.

Lincoln's classes were sped up to lengthen the Christmas break and alleviate congestion on public transportation for traveling service members. Some students reportedly wore pajamas while attending Saturday classes.

NATIONAL DEFENSE TRAINING

Another significant addition to the Lincoln campus was the National Defense Training courses taught by former CPT cadet and Lincoln alumni Walter Sanderson.

Sanderson, who was named to the Lincoln University Alumni Association's Hall of Fame, went on to open his own manufacturing business. He arrived at Lincoln already skilled at metalworking. At age nine, he was sweeping out a machine shop in Fulton, and by age fifteen, he was operating a lathe. He developed fourteen-hour-day work habits, between summers threshing the fields and otherwise metalworking.

These two cannons, which sat outside Memorial Hall facing the quadrangle, became part of the Lincoln University of Missouri scrap metal drive at the beginning of World War II. *Lincoln University Archives.*

Fulton's Walter Sanderson was in the first Lincoln CPT class. During most of the war, he taught National Defense Training vocational education courses in Damel Hall. After the war, he owned and operated a successful manufacturing company. *Lincoln University Archives.*

The eight-week courses were taught from 6:00 p.m. to midnight Monday through Friday. Community members, faculty spouses and other students were taught how to work with a lathe, milling machine, shaper and a welder. In its first eighteen months, the program placed 73 percent of its 162 graduates in defense plants in St. Louis, Kansas City, Seattle, Pittsburgh, Rock Island and California.

"Women in the various classes are quicker and a bit more apt than the men," Sanderson said. Student Fanny Peeples noted that the skills from the Lincoln machine shop gave her a way to do her part in the Allied war effort, other than joining the Women's Army Auxiliary Corps (WAACs) or Women Accepted for Voluntary Emergency Service (WAVES).

The university's new equipment in Damel Hall was directly part of war production as well. "Little is it known, but it is a fact, that the B-29 Superfortresses that have been giving Tokyo a little bombing trouble of late from Saipan bases have parts that were made on the LU campus," the *Clarion* reported in December 1944. The antenna bracket for the radio was built by employees of the Eagan Machine Shop, who used the lathes, drill press and milling machine in the north-end basement almost daily from April 1943 to December 1944.

Chapter 6

INTO SERVICE

By February 1943, the number of Lincolnites in service had reached two hundred. "The vacancies in the men's dormitories are so noticeable that these three buildings are fast assuming the aspects of a deserted village. Dormitories that once were crowded with three students in many rooms can hardly present two to a room. Trunks that once filled the baggage rooms are now few, and many of the students still residing in these buildings are packing now," the *Clarion* reported on February 19, 1943.

Not surprisingly, most of the new students for the spring semester were women. To show their support, these female students began dressing for Thursday night meals and proposed an increase to the number of dances, if not providing a location for the men and women to socialize on campus beyond the library and meals. The men agreed, as one letter to the editor writer put it: "We want to dance here before we are called to dance in Tokyo."

Though not leaving in the same numbers, nearly two dozen Lincoln women would serve in the WAACs, including Jefferson City–born Captain Harriette (Hardin) West. "We're enthusiastic over our membership in a new organization for women which affords so many opportunities," WAACs recruit Ruby Smalls told the *Clarion*.

"To be inducted into the army causes little anxiety in itself, but the mystery as to when the induction will occur is eating every man's thoughts," the first *Clarion* in 1943 said. "Members of the Enlisted Reserve Corps feel the situation as some sort of shadow slowly creeping about them....School work takes on an entirely new aspect. It becomes something that does not really matter, because other problems have forced it to the background. Men find it easy to become care-free, idle and indignant. And it's all because of the uncertainty of it all."

COACH CHARLES HOARD

Lincoln's students were not the only ones to put their future dreams on hold to serve their country.

Charles Hoard, a second-year Tiger line coach and proctor at Yates Hall, chose to enlist rather than return to Lincoln for the spring 1943 semester. He provided physical training to pilot cadets and ground crewmen in basic training at Keesler (Mississippi) Army Air Field and later at Tuskegee Army Air Field.

Born in Mississippi, Hoard was the first of nine siblings to earn advanced college degrees. Hoard first arrived at the Lincoln campus as a student from St. Louis on an agriculture scholarship, paying his way by milking cows, according to his daughter Adrienne Hoard.

Charles Hoard had been an assistant football coach at Lincoln for one season when he enlisted. He provided physical training for the Tuskegee Airmen at Keesler and Tuskegee Army Air Fields. Hoard remained in Jefferson City's U.S. Air Force Reserve no. 9698 after the war, being promoted to major. *Adrienne Hoard Collection.*

Hoard left before graduating to work as a cook for the railroad. Finding himself in Texas, he completed his degree in English at Bishop College, Marshall, Texas, where he remained as dean of men, English instructor and coach. Before returning to Lincoln as a professional, Hoard taught English three years at Bruce High School, Brunswick, Missouri, where he coached the football and basketball Trojans.

Hoard was "a star within his own rights, having set a great record in football and basketball at Lincoln and as a high school coach," said a 1941 Richard Jackson sports column in the *Clarion*, when Hoard was named the Lincoln football team's line coach. He coached many CPT cadets and several who went on to become Tuskegee Airmen.

In the spring of 1943, Hoard was one of nine African Americans at the U.S. Army Air Force Physical Training School in Miami Beach, Florida. Among his six-week cohort were seven All-American football players, several All-American basketball players and boxing title holders. Only those with physical education degrees or experience in physical education were invited to the new training school, which made the trainees eligible for Officer Candidate School. In addition to classes in administration and physical exercises, the trainees swam daily in the Atlantic Ocean. Hoard reported to the *Clarion*, "If I get out okay, I'll take on Joe Louis first and then the Chicago Bears Football team, alone."

Professor Stewart Fulbright

At the same time, French professor Stewart Fulbright Jr. received his call up as a private in the U.S. Army Air Force Reserve on January 31, 1943, into Flight B of the 1166th Technical School Squadron at Jefferson Barracks.

After earning his bachelor's degree in French in June 1941, Fulbright remained on campus as a foreign language professor. His may be the most interesting route of all Lincolnites who found their way to Tuskegee. The Springfield, Missouri native was the last of six Lincolnites to earn wings at Tuskegee.

When he arrived at Jefferson Barracks for the rigorous entrance exams, he was the only Black man in a group of twenty-two that day. Only eight of them passed the written intelligence portion. Only Fulbright and one other candidate also passed the physical exam.

He later told the *New York Age*, "I thought it strange that I was the only Negro in a group of over 2,000 aviation cadets and I later found out that there were 3 of us there by mistake....I can sincerely say that during this processing [mental, physical and orientation tests, interviews and drilling] no discrimination was shown toward me." Fulbright told the *Clarion* staff that his white fellow cadets accepted him in conversation and horseplay; however, he chose to eat in a separate mess hall.

Fulbright was notified nine days after induction that he would be honorably discharged because "the Omaha office [headquarters of the Seventh Area Command] made a mistake." He was allowed to keep his uniform and was still subject to the air reserves.

He resumed his duties as French professor and met his future wife, Della (Cox), a Jefferson City native attending Lincoln High School. Before the semester was over, he was training at Tuskegee Army Air Field.

Stewart Fulbright earned a degree in French from Lincoln and then remained as an instructor. The Springfield native graduated from Tuskegee in December 1943 and was assigned to the 477th Bomber Group. *Lincoln University Archives.*

WEDDING BELLS

For several dozen men on campus, the uncertainty of active duty did not last long. The call came for sixty-one students from the Lincoln Enlisted Reserve Corps (ERC) in March 1943 to report to Jefferson Barracks. That left nineteen still waiting for their news. And another twenty-five recently became eligible for the draft and expected their call before semester's end.

Among the ERC men called up was senior William Randol, who served in the 743rd Military Police (Aviation) with two dozen of his fellow Lincolnites. Before he left campus, the *Clarion* revealed that Randol and his wife, Willie Porter, a Lincoln graduate, had married ten days after the Pearl Harbor attack but had kept the matter a secret for sixteen months, even from their parents. They hoped to reveal the news after Randol's graduation in June 1943. Instead, they traveled to Chicago and Cape Girardeau to tell their respective parents before his induction.

Several other Lincoln couples followed suit in the few weeks between news of their call up and the date of their induction, including fellow 743rd recruit and CPT cadet Bertran Wallace. He married mechanical arts department employee Mildred Black in Linn. Wallace had been the

Cement College Romance

Charles Anderson, a 1941 chemistry graduate, returned to Jefferson City after graduating air corps meteorology school to marry Marjorie (Robinson), before reporting as a weather officer at Tuskegee Army Air Field in May 1943. *From the* Lincoln Clarion, *May 21, 1943.*

supervisor in the graphic arts department and was expecting to graduate in June. At the time of his call up, he also was senior class president and vice-president of Alpha Phi Alpha chapter.

While still at Camp Whiteside, Fort Riley, Kansas, several members of the 743rd Military (Aviation) Police visited the campus in May 1943. Wallace, Randol and Maurice Harris and their new brides were among ten newlywed couples honored with a reception by the campus faculty and staff. They were joined by CPT cadets Walter Sanderson and Robert Taylor and their wives for a "Recorded Music Evening" at Anthony Hall.

In anticipation of being called to service in December 1942, Maurice Harris and Lucille Buchanan, who was the secretary in the agriculture department, were the first couple married in Page Auditorium in its forty-nine-year history. Fulbright and Hoard were ushers, and her bridesmaid was the sister of Tuskegee Airman Richard Pullam.

Also in May 1943, the 1941 chemistry graduate Charles Anderson returned in Jefferson City after graduating from the U.S. Army Air Corps' Institute of Meteorology at Chicago University. The weekend before reporting for duty at the Tuskegee Army Air Field, as one of only five Black weather officers in the U.S. Army Air Corps, Anderson married his college sweetheart, Marjorie (Robinson), daughter of local community leader and baseball legend Charles "Lefty" Robinson.

FAREWELL

President Scruggs organized a special convocation in March 1943 to honor the more than two hundred men in service, the sixty-one on their way and the remaining men waiting for their call.

Half of the men called in March 1943 from "The Hill" were assigned together to Company A of the Army Air Corps' 743rd Military Police

Drummer Floyd Ramey was the only Jefferson City native who took the Civil Pilot Training program at Lincoln University of Missouri and then joined the U.S. Army Air Corps. *Lincoln University Archives.*

Battalion and sent to Fort Riley, Kansas. The combination of their athletics and academics made this company a well-known novelty once the men reached Europe.

This military exodus also included several pilot hopefuls headed to Tuskegee, like CPT cadets Wilbur George Jr. and Louis Harris. Edwin Barrett and Victor Barker, ERC members, received deferments just after surviving an early Hell Week to become Alpha Phi Alpha members.

CPT cadet Floyd "Jelly" Ramey, a Jefferson City native, also left for Keesler Army Air Field. "[I'm] back at my old job playing in the orchestra, only this time, I am not playing drums, I am playing the bass fiddle and I like it," he reported back to his alma mater. By the fall, he had deployed to England with an aviation quartermaster company. After the war, Ramey returned to his hometown, serving at the Toney Jenkins Post of the American Legion.

After the March 1943 loss of men to service, campus life adjusted. Fraternities had sped up the pledge process before the upperclassmen left. The campus band postponed its spring concert. The journalism department opened to sophomores, more women filled leadership roles on the *Clarion* staff and the print editions were reduced to four pages.

The Stagecrafters had to postpone the latest play, which included student Jimmy Stewart in a lead role. His poem "Red River" was published in the *Clarion* after being read at the March 1943 convocation for servicemen, and it later won a national poetry contest. Stewart became Lincoln's third alumni to be killed in action.

No Black Navy Pilots

In April 1943, the U.S. Navy reminded CPT cadet Earl Hogan Jr. that it was still a segregated military branch. The only role available to Black sailors for two decades prior to December 1941 was mess attendant. Although the military branch expanded roles during the war, they remained noncombat.

A U.S. Navy recruitment advertisement invited readers to request pictures of planes be sent to them. The reader was encouraged to identify what type of airplane each picture represented as a test of their aviation knowledge. Hogan's answers must have impressed the service branch.

He was "greatly interested in fighter and bomber planes," the *Clarion* reported after Hogan applied. The White Plains, New York naval branch sent Hogan the St. Louis recruiter's contact information. And the navy sent the Pilot Grove native a questionnaire for placement in naval aviation.

Upon receipt of his application, which indicated his race, the Naval Aviation Cadet Selection Board at St. Louis sent a reply. The board appreciated his application but couldn't process it, recommending that he enlist in a different role with the U.S. Navy. The *Clarion* observed, "If the color of a man's skin doesn't matter while the man is at the controls of a plane, then why does it matter when a man seeks placement in naval aviation?"

A graphic arts major, Hogan came to Lincoln in 1940 from Texas. His colorful past included dancing with Count Basie and serving as prop man with Harlan Leonard's band. He was known for dressing sharply and being a "most enterprising 'hep cat.'" On campus, Hogan played drums with the Collegians and was a formidable flyweight boxer and star halfback. He also was involved in the intramural council and organized wartime physical fitness.

A graphic arts major, Earl Hogan was in the third Lincoln CPT class. He was active and popular on campus. He joined the U.S. Navy in June 1943 and was at the USN Magazine at Port Chicago, California, when the devastating explosion occurred. *Lincoln University Archives.*

Despite the aviation racial barriers, Hogan joined the U.S. Navy two months later, completing petty officer school at Great Lakes, Illinois. Hogan and another Lincolnite, Harold Hardiman, were stationed at Port Chicago, California, at the time of the great disaster on July 17, 1944. Hogan told the *Clarion* that he was "lucky enough to escape without injury." Hundreds of Black sailors were not as lucky.

Two ammunition ships and a resupply train exploded while being reloaded. About five thousand tons of bombs, depth charges and ammunition were being moved between the rail cars and two merchant ships. The effects of the blast were felt a state away, and residents of the nearby town were literally knocked off their feet. More than 320 Black sailors lost their lives, with hundreds more wounded.

The site was so obliterated that an exact cause leading to the explosion could not be identified. However, the crews of Black sailors had not been properly trained in handling the potentially explosive weapons, ammunition and related equipment they were charged with loading and unloading aboard ships. Plus, white superior officers irresponsibly encouraged the men to cut corners to speed up the process.

The Black lives lost in that moment accounted for 15 percent of all African American deaths in service during World War II. After being ordered to clean up the wreckage, the remaining Black sailors were ordered to a new port to resume loading the weapons, ammunition and related equipment without any further training. Nearly 260 surviving Black sailors refused the order, demanding improved safety and training. Most of those men were court-martialed for disobeying orders and given bad conduct discharges, which were later expunged in 1946.

SEMESTER'S END

Before the spring 1943 semester ended in June, the campus was affected by news of George Washington Carver's death, inspired by Lincoln history professor Lorenzo Green having published his *The Negro in Colonial England, 1620–1776* and discouraged by the defeat of a promising Missouri civil rights bill in the General Assembly.

War news brought great pride and great sorrow to the campus. Unexpected national photos of Lieutenant Wendell Pruitt and Lieutenant Richard Pullam as part of the 332nd Fighter Squadron showed that they had been deployed overseas and were involved in military maneuvers over Tunis. It was the first Pullam's brother and sisters on campus had heard of his deployment overseas. Like for the rest of the country, the sight of African American men sporting flight suits and the idea of them soaring through the air was an inspiration to other Black men and women.

The campus also learned that Staff Sergeant Maynard Harris became the second Lincolnite killed in action. Harris was a basketball star for the Lincoln Laboratory High School, from which he graduated in 1940. Formerly a filling station mechanic in Festus, Harris was serving with the 1963rd Ordnance Company (Aviation) when he died in Tunisia in February 1943, after volunteering for service eleven months earlier.

Chapter 7

BECOMING TRAILBLAZERS

Although Black Americans disagreed over whether the U.S. Army Air Corps' 99th Fighter Squadron should have been forced to integrate rather than form as a segregated service, they must have agreed that seeing Black pilots, mechanics and service members exceed white expectations was extremely satisfying.

A segregated service meant that an entire system of training, facilities and command structure had to be built from scratch. The first step was identifying a location. Lincoln Field and Moton Field, one private and the other used by the local institute, were the first sites to become part of the now legendary Tuskegee Army Air Field.

While construction began on the military facilities in Alabama in early 1941, the first class of ground crewmen arrived at Chanute Field, Illinois, where white airmen were training for the same support and technical roles. Among these first 250 Tuskegee Airmen, whose training uniquely was integrated, were Clovis Bordeaux and Robert White. They became mechanics, aircraft armorers, aircraft supply clerks, technical clerks and instrument mechanics.

Opposed to Black men in the U.S. Army Air Corps, General Henry Arnold, in charge of the new U.S. Army Air Forces, created in June 1941, said, "Black youth could perhaps be trained as pilots in 18 months, but it would take years to train enlisted men as mechanics, especially keeping within the framework of the separate-but-equal doctrine."

Instead, this first class of Black enlisted men completed their non-flight training in eight months with the "most talented scores to graduate Chanute." Major General Rush Lincoln, chief of the air corps technical training command, noted the 99th Pursuit Squadron ground crew's exceptional classroom performance: "If operations under test conditions are successful, the colored unit will have been proved sufficiently to warrant extension of this type of training."

Bordeaux and most of his fellow enlisted men had earned an education well beyond high school. This first group of Black enlisted men in the U.S. Army Air Corps were equal to the academic skills and general quality of those who qualified as pilots. "Some of the men had a very difficult time adjusting to their white supervisors, instructors and officers, most of whom weren't nearly as qualified to teach, nor direct," noted the National Park Service.

Options for these Black ground crewmen in the segregated Army Air Corps were limited after they graduated in October 1941. They were assigned either to Moton Field, the primary training field at Tuskegee, or Maxwell Army Air Field, Montgomery, Alabama. At the latter, they were not given duties for their skills or experience and were not in positions to acquire rank, relegated to doing things like housekeeping and landscaping.

At Tuskegee, the remainder of these skilled and intelligent Black men helped complete construction of the Tuskegee Army Air Field.

The military construction contract for Tuskegee Army Air Field was the first awarded to a Black-owned firm, McKissack and McKissack, which converted a wooden graveyard into a military air field beginning in July 1941, following the design of African American architect Hillyard Robinson of Washington, D.C.

Nearby Tuskegee Institute had been one of the first Black schools awarded a Civil Pilot Training program contract and was the first to offer the advanced CPT program. That success, plus serious lobbying by school officials, caused the military leadership to establish the segregated air base near the school.

But the nearby town of Tuskegee was not fond of the influx of so many Black soldiers, particularly officers and especially those who carried weapons. The racist politics of the town had an influence on base, such as when Bordeaux served on guard duty without bullets. Several instances set the white townspeople against the Black military police, and the military leadership continually sided with the whites, asking the Black servicemen, who had done nothing wrong, to let it go.

For example, white residents of Tuskegee attacked Black military police sent from the base to escort military personnel back to the base from the local police department. Seeing Black men carrying firearms in their town, despite it being in their professional capacity, the white mob beat up one of the MPs severely. The disappointing outcome was that Black MPs would no longer carry sidearms off base.

FLIGHT TRAINING

The first pilots of the 99th Fighter Squadron began with five weeks of preflight training. Some ground school subjects—such as flight theory, engines and navigation—and army regulations resembled the CPT program. Ground school for pilots at Tuskegee also included Morse code, Link training and flight simulations, aerodynamics, propellers and meteorology.

Cadets first took to the air in primary training at Moton Field in the two-seat open cockpit of a Stearman PT-17 "Kaydet." Moton Field was built by the school for its CPT courses. The primary flight school mechanics and auxiliary staff, early on, were civilians, including women, like Marjorie Cheatham, a licensed and certified aviation mechanic.

Flight instructors initially came from the Tuskegee Institute Civilian Pilot Training program, including Charles "Chief" Anderson, who eventually became the chief instructor. Chief Anderson (not to be confused with Lincoln's Charles E. Anderson, meteorologist) was called the "grandfather of Black aviation." He taught himself to fly in 1926 after buying his own plane when no instructor would teach him. Eventually, he became the first African American to earn a transport license.

Only one of four future runways had been built, but not paved, when the first cadets of the 99th began basic flight training at Tuskegee Army Air Field. Tents served as communication center, cadet ready room and other necessary facilities. The first enlisted men and pilot cadets alike lived in this "tent city." For basic, they flew BT-13 Valiants, a low-wing, 450-horsepower monoplane with a sliding canopy.

The advanced flight training in a North American T-6 "Texan" was conducted exclusively by white instructors. Some were impartial, but most reflected the racial prejudices of the day, which led to many Black pilots washing out of the flight program.

When the first class graduated from Tuskegee flight training in March 1942, only five of the original thirteen received their wings. Among the first five graduates was Captain Benjamin O. Davis Jr., the fourth Black graduate of West Point (1936), formerly serving in the regular army under his father, Brigadier General Benjamin O. Davis Sr., at Fort Riley, Kansas. The younger Davis was a fierce commander of and advocate for the Tuskegee Airmen. He was handpicked by army leadership to eventually command the Black fighter squadron, although he was not a natural flyer himself.

FIRST LINCOLNITES AT TUSKEGEE

Nearly thirty Lincolnites followed Robert White and Clovis Bordeaux through training and service at Tuskegee Army Air Field and Keesler Field, Mississippi. Even more also served in the segregated U.S. Army Air Corps at other bases. All of them met prejudice at every turn.

One month after the historic graduation of the first Black military aviators, the first pilot hopefuls from Lincoln arrived. Asbury Garry Jr., Richard Pullam and Wendell Pruitt enlisted at Chicago, Fort Riley and Tuskegee, respectively. But all began preflight training in April 1942.

The son of a pipe cutter, Garry was born in South Carolina but graduated from Evanston, Illinois, before arriving at Lincoln. He was a shipping clerk in Chicago when he enlisted. He washed out of the pilot training but remained in the air corps for three years. After the war, he owned a grocery store in Evanston, Illinois.

Pullam and Pruitt, who interestingly were in the same Lincoln CPT class, were the first of Lincoln's six Tuskegee Airmen pilots to earn their wings in class SE-42K (single engine, 1942, November). The second lieutenants achieved a rare honor for that time. In the ninth class of Black military aviators, they were among only one thousand Black men to complete U.S. Army Air Corps pilot training during the Second World War.

Pullam's father, Arthur, played semi-pro baseball in Kansas City under the nickname "Chick" and was a postal worker. Pullam was a twenty-three-year-old sophomore when he took the CPT course and worked as a railroad chair car attendant to pay for his education. He was an athlete like his father and joined fraternity brother Bertran Wallace on the Lincoln football field.

Pruitt was the seventh child of a clothing salesman in St. Louis. Like Wilbur Long, he transferred to Lincoln from Harris Teacher College when he learned of the CPT program, said his sister Mrs. V.C. Payne. Pruitt was involved in the drama club Stagecrafters and student government and worked at the Blue Tiger Café, a mainstay of campus life.

Everett Bratcher joined the growing number of Lincolnites at Tuskegee in May 1942 but did not begin his preflight training until December. He was only a sophomore at Lincoln, but he had been in its first CPT class. A building engineering major, he paid for his education as a mail clerk at the Missouri Workmen's Compensation office. Born in Arkansas to a schoolteacher, Bratcher grew up in Poplar Bluff and enlisted at Fort Leonard Wood. During his time at Tuskegee, he was continually described as having excellent character and being a satisfactory soldier. He continued in the air force after the war, and his headstone at the Jefferson Barracks National Cemetery, St. Louis, echoes him having "The quiet dignity of excellence."

Instruction at the Tuskegee program, built from scratch, continued, despite the lack of trained African American personnel, planes, parts and instructors.

FALL OF 1942

Bordeaux and White had just returned with the 99th from gunnery training in Florida when Secretary of War Henry Stimson visited Tuskegee Army Air Field for the first time in October 1942. He "highly praised" the 99th Fighter Squadron and indicated that the squadron would go abroad soon.

Sedalia mechanic Hinton Parker arrived at Tuskegee Army Air Field in November 1942. Parker was married to Clara, daughter of Jefferson City's James Mayberry, who operated the Chicken Coop, a popular restaurant across the street from Lincoln campus.

The racial tension at the Tuskegee Army Air Field changed dramatically in December 1942 when Lieutenant Colonel Noel Parrish became commanding officer of the Tuskegee Advanced Flying School, after serving as director of training from its inception. Base facilities, like the PX and the meal hall, were desegregated under his command. He was also an advocate for equal opportunity.

About this time, the 332nd Fighter Group was activated with the motto "Spitfire" and call sign "Percy," later changed to "Subsoil." The 332nd comprised three fighter squadrons—100th, 301st and 302nd. For flying hopefuls, the expansion meant more seats for fighter pilots. For those who washed out, it meant more ground and support roles.

After graduating in December 1942, Pullam and Pruitt were assigned to different squadrons. Pullam, the first Black pilot from Kansas City, served with the 100th Fighter Squadron. Pruitt was the second Black pilot from

St. Louis, following Lieutenant James McCullen, and was assigned to the 302nd. Although Pruitt's squadron was "considered the least desirable," according to Tuskegee Airman Charles Francis, they developed a strong sense of team spirit.

In January 1943, while waiting for its deployment, the 99th lost one-time Lincoln CPT instructor Lieutenant Richard Davis as its second casualty. Davis left Lincoln for Tuskegee, where he was among the first thirty pilots to earn their wings. He died in a training exercise.

SERGEANT WINSTON ROGERS

Lincoln CPT cadet Winston Rogers arrived at Tuskegee as a primary flight instructor in January 1943. He was soon assigned to Scott Field, Illinois, to complete cross-country and instructors ratings. Then he was trained in radio operation and mechanics at the air force's technical school in Tomah, Wisconsin. He also was assigned to MacDill Army Air Field, Florida; Maxwell Army Air Field, Alabama; and Turner Army Air Field, Georgia.

Winston Rogers was an Olympic-level track star when he arrived at Lincoln. After completing the second Lincoln CPT class, he earned his commercial pilot's license and then became a flight instructor at Chanute Field, Illinois. After being drafted, he was transferred to Tuskegee Army Air Field as a flight instructor. *Lincoln University Archives.*

Still hoping to become a commissioned army pilot himself, Rogers drafted many letters to the U.S. Army Air Corps seeking placement. "I know I can be of more service in aviation to my country than as a ground crewman....If you would give a chance, I'm sure I could prove my ability; but without I have no way of using my talent for my country," Rogers wrote.

Born in Independence, Kansas, Rogers was an outstanding track athlete and came to Lincoln as a PE major. A "long-limbed youngster, [Rogers'] record-smashing leaps were emblazoned on the sports pages of the nation's newspapers while he was still in high school," the *Pittsburgh Courier* said. In the fall of 1941, Rogers lost a meet by half an inch to the world record holder, Lester Steers.

When he did not initially pass the U.S. Army Air Corps exams, he pursued a commercial pilot's license and then became a flight instructor at Chanute Flying School, Illinois, with more than two hundred flight hours. "Flying has been in the soldier's blood ever since the bug hit him at Lincoln," the *Clarion* said.

Rogers was drafted into the 84th Aviation Squadron, the first Black squadron activated at the U.S. Army Air Forces Basic Training Center, Jefferson Barracks, in the fall of 1942. He was the acting supply sergeant when his letter arrived to be a primary flight instructor at Tuskegee.

Discharged in 1946, he reenlisted with the U.S. Army Air Corps as a tech sergeant and was deployed to the Philippines as a company administrator. In 1948, he reenlisted again with the new, desegregated U.S. Air Force Reserve as an instructor. As late as 1949, Rogers still was seeking an officer's commission, for which he was denied "due to lack of ground combat experience."

Rogers eventually completed his education degree at Pittsburg (Kansas) Teachers College. The school brochure said that Rogers "excels at sports, track, baseball, football, basketball, softball, tennis, boxing and wrestling." He went on to earn his master's degree in education administration.

To Europe, to Selfridge Army Air Field

In March 1943, the 332nd Fighter Group was reassigned to Selfridge Army Air Field near Oscoda, Wisconsin. A month earlier, fellow Lincolnite Clarence "Ozzie" Farquhar joined the 332nd at Tuskegee Army Air Field after his technical training at Chanute, Illinois, with the warrant officers. The son of an Illinois coal miner, Farquhar completed one year at Lincoln before enlistment, when he was working as a monument apprentice. Farquhar was

Major Richard Pullam and Captain Wendell Pruitt uniquely earned their commissions at Tuskegee Army Air Field in the same class, 42K, December 1942. Pullam's wings were pinned by his brother, Arthur Pullam Jr., who was a longtime biology professor at Lincoln. *Lincoln University Archives.*

assigned as mail clerk to the 100th Fighter Squadron, when he wrote to the *Clarion*, "I am beginning to like life with the air corps."

Before rejoining the 332nd in Michigan, Pullam and Pruitt stopped in Missouri to visit their homes and alma mater. Pullam visited his fiancée, Carolyn White, who was due to graduate in June, as well as his sisters, Elisabeth and Barbara, who were Lincoln students, and his brother Arthur, a biology professor. Pruitt and Pullam were guests of their "third musketeer," Bertran Wallace, who remained a campus leader and athlete after racism had diverted his pursuit of military aviation. Then they visited Pullam's family in Kansas City and Pruitt's in St. Louis.

While the relocation of the 332nd and the deployment of the 99th overseas alleviated some congestion on the small Tuskegee Army Air Field, being the only segregated base for Black U.S. Army Air Corps members, Tuskegee was continually filled beyond capacity.

LARGEST LINCOLN CONTINGENT AT TUSKEGEE

The highest number of Lincolnites at Tuskegee at the same time likely was the following month, in May 1943. Asbury Garry probably was the senior Lincolnite on base, having been there more than a year. Everett Bratcher was in the midst of basic flight training, James Merritt and Milton Lemmons were in primary flight training and Louis Harris, Edwin Barrett and Wilbur George were in preflight training.

Barrett wrote back to the *Clarion* that their studies included physics, math, geography and military courses. "We arise at 5 in the morning and go constantly until 10 at night. We must be on the ball every minute of the day—shoes shined 2 and 3 times a day and our rooms kept spotlessly clean at all times. We go to all places in formation and must observe the rules of etiquette at all times. We eat with one hand under the table, etc. In other words, they train us to be gentlemen and officers, then pilots," he wrote.

New to the segregated air base in May 1943 were Charles E. Anderson, Stewart Fulbright, Curtis Green and Junius Savage.

Four more Lincolnites arrived in May 1943 at nearby Keesler Field, a U.S. Army Air Forces technical training school. Lincoln football coach Charles Hoard was a physical training instructor, and Lewyn Boler, Wilbur Long and Floyd Ramey were recruits training to move over to Tuskegee. That spring, Tuskegee Army Air Field hosted a twenty-three-team athletic meet, where Hoard's Keesler field team took third place.

Meteorologist Charles Anderson

Lincolnite and May arrival Charles E. Anderson (not to be confused with the acclaimed pilot "Chief" Anderson) completed the Army Air Forces Meteorological Aviation Cadet Program at the University of Chicago. "It was a rigorous program that pushed young cadets, weekly tests in each subject were administered and only those that passed could continue on."

Growing up on a farm in University City, he was valedictorian of Sumner High School, St. Louis, in 1937. At Lincoln, Anderson, who went by "Sags," was president of his pledge class and his fraternity, as well as president of his class for three years. He also coauthored the 1941 yearbook's seventy-fifth anniversary tribute to the school's history. He was among the first and few Black weather officers. His role was to teach Black pilots how to predict the weather.

Anderson was assigned to Selfridge Army Air Field in January 1944 soon after the 553rd Fighter Squadron was organized there to train replacement pilots for the 332nd Fighter Group in Europe. Anderson likely was on base when the attempt of Black officers to enter the segregated Lufberry Hall incident occurred, resulting in the 553rd being reassigned to Walterboro Army Air Base, South Carolina, in May 1944. The fighter squadron was soon disbanded, and Anderson joined the 477th Bomber Group at Godman Field, Kentucky, in October 1945.

Everett Bratcher

Poplar Bluff native Everett Bratcher was the third Lincoln CPT cadet to earn his commission on August 30, 1943, at Tuskegee Army Air Field with class SE-43H (single engine, 1943, August). Before getting his call to Tuskegee, Bratcher worked at the Missouri Workmen's Comp office in Jefferson City.

After graduation at Tuskegee, Bratcher was assigned to the 301st fighter squadron at Selfridge Army Air Field, Michigan, where Pruitt and Pullam just had been promoted first lieutenants.

The 332nd Fighter Group in Michigan was training to achieve "A-1 efficiency in flying, shooting, mechanics and theory, so that they will be ready to take their places beside flying comrades already in the thick of the battle on fighting fronts," the *Courier* reported. While Pullam and Pruitt received their orders to deploy overseas, Bratcher learned that he was disqualified

Everett Bratcher did not deploy overseas with the 301st Fighter Squadron but rather served as armament and chemical officer at bases in Tuskegee, South Carolina and Illinois. After the war, Bratcher worked thirty-five years with the Defense Mapping Agency as a cartographer. *Missouri History Society Everett Bratcher Collection.*

from overseas duties. Bratcher served three years as armament and chemical officer, supervising air and ground crews at Tuskegee, Alabama; Walterboro, South Carolina; and Scott Field, Illinois.

As did many Black pilots and aviation mechanics during the war, Bratcher hoped to make a civilian career out of aviation. He attempted to work as a commercial pilot and traffic controller but found that racist mythology continued to block his path. Instead, he worked thirty-five years as a civilian cartographer for the Defense Mapping Agency. He received several recognitions for his contributions there, including for "sustained superior performance" and for improving efficiencies. Bratcher continued flying for himself with the Scott Air Force Flying Club.

Stewart Fulbright

Tuskegee graduated its second class of twin-engine pilots in December 1943; among them was Stewart Fulbright, with class TE-43K (twin engine, 1943, November). Fulbright and the other twin-engine graduates were sent to Mather Field, California, for training with the B-25 Mitchell bombers in January 1944 with the reinstated 477th Bomber Squadron.

Between his graduation and his new assignment on the West Coast, Fulbright stopped in Missouri to marry Jefferson City native Della Cox, who at the time was a junior commercial education major at Lincoln.

When Fulbright arrived at Tuskegee, all he knew was that he would learn to fly and that he hoped he was in the half of the cadets in the end that did not wash out, he said in an oral history interview. In the earliest training, Fulbright and his fellow cadets rode a bus from Tuskegee Institute dormitories to Moton Field, where Black civilian instructors encouraged the young hopefuls. The ground school, however, was severely rigorous, with a variety of subjects to be mastered, such as weather, propellers, engineering and combat, Fulbright recalled. Each cadet in his class specialized in an area—for Fulbright it was weather—and then they helped one another earn the best average scores for any class in the Southeast Training Command, he recalled.

Fulbright said that the men in his class grew close because "we were all going through the same thing, we all had the same goals; you learned early on to support one another." When a classmate was taken off flying status, everyone was disappointed, Fulbright said.

The greatest lesson Fulbright gained at training camp was the value of teamwork: "When there's a serious problem or great adversity, you have knowledge, clear goals and figure out how to confront it—together." The lesson of teamwork applied as much to the segregated environment as it did to gaining the aviation skills.

Some white instructors were clearly racist and not only made the Black pilots' lives difficult but also had the authority to remove them from the flight program. "We were on pins and needles whether it would be our last day, every day," Fulbright said. But other white instructors were friendly and encouraging, he noted.

The upperclassmen had their own rituals of hazing the underclassmen, like recitation of nonsense poems and eating a "square meal" with the fork drawn straight up from the plate then straight to the mouth. "I had been through fraternity initiation at college—this was quite different," Fulbright said.

END OF 1943

Other Lincolnites at Tuskegee at the end of 1943 included Milton Lemmons, who completed preflight training and was assigned as a sergeant in the support crew. Edwin Barrett, Wilbur George Jr., Louis Harris and L. Wright all were in the same lower primary class, while Curtis Green was in basic flight and Wilbur Long in upper basic. James Merritt was just in preflight training, and Victor Barker was in the upper preflight.

Harris updated the *Clarion*: "We, the Lincolnites at the Cadet Detachment at Tuskegee Army Air Field, have found the *Clarion* quite a source of enjoyment."

Another Lincoln CPT cadet, Glenwood Johnson, was called up from the AAC enlisted reserve, arriving at Keesler Field in November 1943. "Sgt. Hoard has been very helpful to acquaint me with army life," Johnson told the *Clarion*. At Keesler, Johnson joined Howdy Green, Chuck Harris, Jewell Shannon, Billy Parris and other Lincolnites.

EARLY 1944

Pruitt, Pullam and Farquhar deployed overseas from Michigan with the 332nd Fighter Group in early 1944. Finis Holt was transferred from Chanute Field, Illinois, to Selfridge, where he was joined by Lincoln's first CPT cadet to fly solo, Wilbur Long, who received his commission at Tuskegee on February 8, 1944, with class SE-44B (single engine, 1944, February). In April 1944, Long and Holt were deployed to Italy as replacements for the 99th Fighter Squadron and ground crew.

That same month, Lieutenant Charles Hoard was reassigned from Keesler Field to Tuskegee after earning his officer's commission in Miami Beach, Florida. Just before that, Hoard was able to serve at Keesler Field at the same time as his brother, Dewitt, who later attended Lincoln. While Hoard may be remembered by his decades of former students as a stern leader with decorum and high principles, in his younger years, his scrapbook reveals that he was a ladies' man, as his daughter Adrienne noted. But time changed him, after Hoard lost five of his closest friends while in service.

LOUIS HARRIS

Fewer than one thousand Tuskegee airmen became pilots; of those, about six hundred never deployed overseas.

For example, Louis Harris nearly earned his wings until he was eliminated from pilot training for "carelessness and lack of judgment" when he neglected to put down the landing gear during a night flight in blackout conditions.

Harris was highly popular on the Lincoln campus and came to Tuskegee through the Enlisted Reserve. A former editor for the *Clarion*, he sent many letters back keeping the campus informed of the progress of its many Lincoln airmen. For example, he wrote soon after his arrival at Tuskegee, "I'm enjoying army life, more or less, in the air corps at this God-forsaken place. Don't let the south get you. Really, though, things aren't so bad."

Louis Harris washed out just before graduation from Tuskegee in November 1944. He was in the fourth Lincoln CPT class. *Daniel Haulman Collection.*

Born in Alabama, Harris grew up in Ohio the son of a foundry worker and teacher. He came to Lincoln after graduating in St. Louis. After the war, he became a probation officer in Pasadena, California.

EDWIN BARRETT

Similarly, Edwin Morris Barrett began his flight training in November 1943 with class SE 44-E (single engine, 1944, May). Unfortunately, just before graduation, he was eliminated "due to flying deficiency." His military records were "very satisfactory," and he was fully qualified medically and academically, his records said. But he was unable to land consistently, and his air judgement was deemed unsatisfactory.

Barrett was born in Muskogee, Oklahoma, the son of a motor mechanic. The CPT cadet majored in business administration at Lincoln, participating in track and football, playing trumpet in the band and singing bass in the choir. He qualified for the U.S. Army Air Corps Enlisted Reserves in 1942, receiving a deferment until 1943.

Learning the motor mechanic trade from his father, Barrett worked for Missouri Supreme Court judge Ernest Moss Tipton while at Lincoln. The judge described Barrett as "honest, trustworthy and [with] unquestionable integrity" in a recommendation letter.

After washing out, Barrett was assigned as a clerk typist at Shaw Air Force Base, Sumter, South Carolina. During his time there, he completed his degree at Allen University in Columbia, South Carolina. He served at bases in Connecticut and Virginia before discharge in January 1946. He reenlisted and served until April 1963 as a clerk. As a civilian, Barrett worked for the U.S. Postal Service. His son, Edwin Jr., graduated from Laughlin, Texas, Air Force Base, and became the first African American pilot for the Georgia Air National Guard.

CURTIS GREEN

A 1940 Lincoln graduate, Curtis Green was no stranger to airplanes and air fields, as his father was the owner and manager of the Lockwood field in Moberly. He completed his primary flight training at Tuskegee in September 1943 but was eliminated from upper primary a few months later. "My coordination was a little too bad for the army to take a chance on my flying a fighter or bomber," he wrote to the *Clarion*. However, Green had enough flight hours to become a civilian pilot after the war.

He was a clerk in Assam, India, for two years. After his service, he worked as a baker in Moberly, where he also made some blues recordings.

CLIFFORD "JEWEL" SHANNON

Another Lincolnite, Clifford "Jewel" Shannon, completed only elementary flight training in September 1944. Then he was transferred to Drew Field, Tampa, Florida, where he was a personnel clerk at the post exchange. He was deployed to Guam with the 1964th engineer aviation battalion in June 1945 and was discharged as a sergeant in February 1946.

Born in Memphis, Shannon lived with his grandparents in Colorado after his father's death. At Lincoln, the chemistry major was an integral part of

Moberly's Curtis Green grew up around the Lockwood Airport, which his father owned and managed. He washed out of Tuskegee flight training due to poor coordination but became a civilian pilot after the war. *Lincoln University Archives.*

78

the Collegians swing band in the trumpet section. He enlisted in September 1942 and first was assigned as bugler with the 48th Aviation Squadron at Kearns Army Air Base near Salt Lake City, Utah. There he had the chance to play with Jimmy Blanton and Wendell Marshall, formerly of Lionel Hampton's band.

After the war, he worked for the U.S. Army Corps of Engineers in St. Louis, earning recognition for "superior performance."

MILTON LEMMONS

Milton Lemmons was eliminated from pilot training while still at Moton Field in May 1943, after his plane drifted into another parked plane during a maneuver. The inspection determined that it was due to poor technique and judgement error. He remained in the air corps as a gunnery sergeant.

The CPT cadet was involved in Alpha Phi Alpha and the Booster Club on campus and completed his biology major at Lincoln after the war. Born in Alabama, Lemmons grew up in St. Louis, the son of an elevator operator and janitor.

Lemmons is one of three Lincolnites included in the St. Louis Science Center's 1992 photo exhibit honoring Tuskegee Airmen. And he is one of four Lincolnites featured among the seventy-five Black aviators in the *Black Americans in Flight* mural at Lambert Airport, St. Louis.

Milton Lemmons washed out of pilot training at Tuskegee Army Air Field and then served on the ground crew. Lemmons is one of three Lincolnites recognized in the *Black Americans in Flight* mural at St. Louis Lambert International Airport. *Lincoln University Archives.*

WILBUR GEORGE JR.

Another Lincoln CPT cadet, Wilbur Isaac George Jr., came from Evanston, Illinois, the son of private servants. During his two years at Lincoln, he was involved in Stagecrafters and Alpha Phi Alpha. Although he qualified for the enlisted reserves, he was called up a year later in March 1943, along with more than sixty other Lincolnites. He did not make it through pilot training at Tuskegee, instead serving at Turner Field, Georgia, until his discharge as a sergeant in early 1946.

JAMES MERRITT

After graduating in 1936 from Lincoln with a math degree, James Merritt hoped to become an electrical engineer. He first attended Lincoln University High School. While at Lincoln, he was on the *Collegian*'s staff, an honor student and a member of the Storks Club and was well-liked on campus.

He was part of Lincoln's first CPT class while working as chauffeur to druggist G.A. Fischer. Merritt enlisted in March 1943 and began preflight training in the fall at Tuskegee. Merritt did not earn his wings. Instead, he served in the support services.

After the war, Merritt was a postal clerk in Kansas City. He is buried at Jefferson City National Cemetery.

James Merritt graduated in 1936 and was in the inaugural Lincoln CPT class. He washed out of pilot training and served in the ground crew. *Lincoln University Archives.*

Chapter 8

99TH FIGHTER SQUADRON

Clovis Bordeaux and Wilbur Long

O f the nearly thirty men from Lincoln who became Tuskegee Airmen, six had the distinction of serving in combat in Africa and Europe. The first to deploy overseas was Sergeant Clovis Bordeaux.

The 99th Fighter Squadron deployed with four hundred ground crew, thirty-three pilots and twenty-seven planes. Bordeaux was among the original ground crew trained at Chanute Field, Illinois, in early 1941. When he arrived for ground crew training, he already held a bachelor's degree in chemistry from Lincoln and a radio operator's license from Milwaukee School of Engineering.

Although he hated flying, Bordeaux had trouble finding a job equal to his skills due to racial barriers. A steady paycheck in the military came at a time when he was still helping to support his widowed mother and younger siblings in St. Louis.

Bordeaux was trained as a communications technician because he already held a first-class radio license. The other equally intelligent non-flight trainees were trained as mechanics, aircraft armorers, aircraft supply clerks, technical clerks and instrument mechanics. Their classes were integrated with the white air corps enlisted men, but they ate and slept separately.

They succeeded where white leadership expected them to fail. The early expansion of the Tuskegee Airmen program is due in part to this group's excellence. They graduated in October 1941 with the highest scores seen at Chanute, when the air corps leadership predicted that Black men couldn't pick up the skills within eighteen months.

Clovis Bordeaux was the first known Lincolnite to join the U.S. Army Air Corps. He was a communications chief for the 99th Fighter Squadron in Europe. *Lincoln University Archives.*

Bordeaux had been promoted to technical sergeant, making him responsible for the planes being ready to fly. "He was a stickler for detail, his entire life...he knew the specs, read the directions. He was very precise in his language, his profession and his parenting," Bordeaux's son, Jacques, said.

Precision was born into his character. His was a "family of folks not only pretty intelligent naturally, but [who] also understood, with their small size, wit and strength of mind got them over," the son said. The Bordeaux family was gifted in math and science, and "precision is the language of science," Jacques Bordeaux said.

What Bordeaux did not care for was logging the required number of in-flight hours each month. Most of the enlisted men in the air corps looked forward to getting up in the air. But "he hated flips and barrel rolls" on Sunday afternoons, the son said.

LAUNCHED FROM LINCOLN

At age thirteen, Bordeaux was the eldest of seven children when his father died soon after the Great Depression claimed his St. Louis hardware store. Although his mother, Elizabeth (Burris), earned a teaching certificate to provide for her family, no one would hire her because she was "a married woman," despite being a widow. "Clovis carried the burden of failing to provide for his family," Jacques Bordeaux said.

Bordeaux was able to attend Lincoln University of Missouri due to the generosity of an uncle. Bordeaux didn't see himself as an exceptional student, graduating from Sumner High School, St. Louis. Yet in his early science courses at Lincoln University, he was a popular study partner because of his knowledge and ability to translate the facts in understandable ways for others.

A member of the Carver Chemistry Club, Bordeaux also helped reestablish the tennis courts and tournaments on campus. He was a member of Kappa Alpha Psi fraternity and played trombone in the Collegians band.

"He enjoyed wearing suits and associating with other Black folks like him," Jacques Bordeaux said. Not all Lincoln students were from upwardly mobile families, but all saw something better for their future.

DEPLOYMENT REALITY

The 99th Fighter Squadron reached its full strength in July 1942 but, unlike white squadrons, waited another nine months before deployment. It was nearly two years after it had been organized when the 99th arrived in the Mediterranean the spring of 1943. "Though they played a less glamorous role, without a doubt, [the enlisted men] played a major role in the success of the Tuskegee [airmen]," pilot Charles Francis wrote.

When the 99th boarded the train from Tuskegee, none had combat experience. As their mission settled in as reality, they developed a sense of deep comradeship. "I felt an unusually strong bond, a linkage with each and every man of the squadron," pilot Charles Dryden said. That fondness extended to Bordeaux, whom Dryden called "a key NCO."

After nine days at sea, the 99th arrived in Africa. The men were accompanied by a war correspondent, Tom Young Jr., from the *Norfolk Journal and Guide*. They stayed at Camp Shanks twelve days, where they were issued bedrolls, gas masks, .45 automatic pistols and winter clothing. Then, without trucks for transportation, they marched three miles in full field gear to camp outside Casablanca, Morocco.

While the squadron and its support units were in Casablanca, Missouri performer Josephine Baker invited the men of the 99th to a performance that was "a showstopper, really something to write home about," Dryden said.

The 99th also benefited from the observation and instruction of Colonel Philip "Flip" Cochran, a flight combat veteran and the inspiration for the cartoon characters in *Terry and the Pirates* and *Steve Canyon*. Cochran warned the pilots of the 99th that the P-40L Kittyhawk flyers "were going to be among the most courageous pilots in the war" because the German ME-109s and the FW-190s could outrun, outclimb and outdive the P-40s. "We would have to stay and fight simply because we were too slow to be able to run away," Dryden explained.

At Oued N'Ja, Morocco, in May 1943, the 99th's new P-40Ls arrived. It had been a month since anyone had been near a plane, and the men were "anxious to get back to business....Just to have new planes rather than

The 99th Fighter Squadron flew P-40 Kittyhawks when they first arrived in the Mediterranean theater. *Library of Congress.*

hand-me-down relics, such as we had at Tuskegee Army Air Field, was a motivator for us," Dryden said. The 99th celebrated the liberation of Tunisia by participating in a parade before leaving Oued N'Ja for the newly built airdrome at Forjouna, east of Tunis.

The 99th flew its first combat mission on June 2, 1943, and faced enemy attack for the first time seven days later. During the campaign against Pantelleria Island, Italy, June 2–11, the squadron averaged two missions each day, requiring Bordeaux and the ground crew to double their efforts on the ground. The 99th flew four missions on June 15, 1943, alone. Some missions involved taking out enemy gun locations, while others involved flying escort for A-20 and B-25 raids. The success of this campaign made way for the Allied invasion of Sicily.

The ground crew's "job was to keep the plane in the air...to keep everything running smoothly as best we were able. If your plane came in with a miss in the engine, you'd stick your hands right down there into that hot engine and take the plugs out to try and get it back in the air as quickly as you could," recalled Tipton native Jim Shipley, who served with the 301st Fighter Squadron as an aviation mechanic.

The 99th settled in to flying escort missions between Tunisia and Sicily and succeeded in its first air victory on July 2. The squadron received three Distinguished Unit Citations for air support provided during Allied landing operations.

After the invasion of Sicily on July 10, 1943, the 99th was relocated to a base at Licata, Italy, on the Mediterranean coast. Often its missions were to beat down the enemy's infantry front lines. Following the end of the Sicilian campaign, the 99th Fighter Squadron moved north in September 1943 to an Italian base near Battipaglia. "We didn't get a chance to gain victories because we didn't go with the invasion force to Italy. We remained in Sicily, hundreds of miles from the battle zone," 99th pilot Spann Watson recalled.

At this time, Lieutenant Colonel Benjamin Davis was relieved by Major George Roberts, who had been the squadron's operations officer. The commanders of the Northwest African Tactical Air Force reported that the 99th performed poorly in combat and suggested that it should be reassigned from combat to coastal patrol. So, Lieutenant Colonel Davis returned to the United States to defend his flyers. In support, Lieutenant Colonel Philip Cochran, a dive bomb expert himself, praised the 99th as "a collection of born dive bombers."

"While colored America thrills to the feats of our fighter pilots, there are scores of officers and men associated with this squadron who never share the glory, but who labor uncomplainingly night and day to 'keep 'em flying and bring 'em back alive.' The greatest pilots would not be able to perform their dare-devil acts or save themselves from destruction, if they did not enjoy the backing of competent and loyal ground officers and crews, hence the saying that a squadron is only as good as its ground crew," Edgar Rouzeau wrote for the *Pittsburgh Courier* in the fall of 1943.

Bordeaux wrote to his alma mater in October 1943 that he enjoyed a cup of "real American coffee—the kind we used to pass up in the cafeteria.... Life here has not been exciting compared with the accounts that have appeared in magazines of combat zones, but it has been enlightening and educational....Students there are passing up breakfast with eggs, while here I am trading a box of bon-bons, which is a vital portion of my candy rations, for six eggs."

Again, the 99th squadron relocated, this time to Madna, Italy, in November 1943. In support of Field Marshal Bernard Montgomery's Sangro River crossing, the 99th flew nine missions in one day, which meant that the ground crew had to reload new bombs, clean the guns, repair the planes and get them refueled. The 99th was proud of the fact that these were integrated missions, with the 99th flying among the other white squadrons of the 79th Fighter Group.

Then the 99th flew multiple close-air cover missions for the British 8th Army ground troops entering Italy in late December 1943. The following month, the 99th relocated to Capodichino Airfield, near Naples, Italy. On January 27, 1944, a squadron of sixteen fighters from the 99th engaged fifteen enemy FW-190s, which were dive-bombing ships near Anzio. They took down ten German airplanes that day but also lost one of their own.

The 99th supported the white New Zealand force on the ground through

Lieutenant Wilbur Long was the third Lincolnite and Civil Pilot Training graduate to fly combat in Europe. He was one of thirty-two Tuskegee Airmen held as prisoners of war. *Naomi Long Madgett Collection.*

the spring of 1944 as part of Operation Strangle, an allied air assault to prevent supplies reaching German forces in Italy. During this time, Lincoln's Lieutenant Wilbur Long arrived in Italy as a replacement pilot for the 99th.

From Pignataro, Italy, the 99th Fighter Squadron flew its 500th combat mission on June 2, 1944, exactly one year after its first combat mission.

In advance of D-Day, June 6, 1944, the 99th had been supporting ground troops. It flew thirty-one sorties on May 11 and twenty-eight on May 12, which contributed to General Mark Clark and the 5th Army capturing Monte Cassino and thereby taking Rome.

Later in June, the 99th transferred to the Ramitelli Air Field, where the 332nd Fighter Group was already stationed. The 99th exchanged its P-40 Kittyhawks for P-51 Mustangs, and it returned to segregated service, no longer flying missions alongside white fighter squadrons. When the 99th was added to the 332nd Fighter Group, it became the only fighter group with four squadrons.

Lincolnites in Europe

Bordeaux and Long would have met up with fellow Lincolnites in-country with the 332nd, including Wendell Pruitt, Richard Pullam, Ozzie Farquhar and Finis Holt. The fighter pilots rotated home after completing fifty to seventy missions, but the ground crew, like Bordeaux, remained throughout the war.

In between missions or in the evening, the enlisted men of the 99th sought diversions. Sergeant Bordeaux, being "an improvisational genius," put his free time, skill and opportunity to work overseas. Officers clubs were standard on air bases, but similar places for enlisted men to congregate had to be made. A self-made "impresario," Bordeaux recycled the wooden bomb crates into walls, and the enlisted men's club was born. In his makeshift gathering place, Bordeaux was well respected. Being neither a drinker nor a gambler, that meant he "held the kiddie" during the big games. This arrangement afforded him enough means to live on, so he could send his entire military paycheck back home, his son said.

Bordeaux also discovered an interest in photography while stationed abroad. He bought up any photo equipment he could find in the local stores, taught himself the art and built his own darkroom out of the same wooden crates. "Most of his life, Dad didn't say much about his service—when asked what he did in World War II, he would say 'nothing…lots of hurry up and wait, even in Africa and Italy,'" Bordeaux's son said.

After his four years of service were up, Bordeaux was sent stateside, before the 99th "really got into the action," the sergeant told his son. "When his commanding officer said he was going home, he was happy to get out of the military, but worried about the loss of a paycheck again," the son said.

Bordeaux received the European Theater Service Medal with eight Bronze Stars, having participated in the campaigns of Sicily, Naples-Foggia, Rome-Arno, Southern France, North Apennines, Rhineland, Po Valley and the Balkans.

Lincoln's Only Known POW Airman

Lincoln's other member of the 99th Fighter Squadron had a very different experience. After flying fifteen combat missions in his P-51, Lieutenant Wilbur Long became one of thirty-three Tuskegee Airmen prisoners of war. Long was flying an escort mission for B-24 and B-17 bombers to the

North Oil Refinery in Blechhammer, Germany, 150 miles east of Berlin, on September 13, 1944. While over the target, Long's Mustang was hit by flak, damaging his canopy and cooling system and injuring him. Thinking that his plane was not too badly damaged, he attempted to make it back to base.

"Several members of my flight warned me that my ship was damaged severely and I should leave my ship," Long said. At one point, he was so disoriented that his wingman, Lieutenant John Daniels, had to turn him back on course from flying in the wrong direction.

When his engine began to freeze up and black smoke was trailing from the right underside of his engine, Long prepared to bail out by parachute. The damaged canopy prevented it. Daniels told Long to find an unobstructed field to land in, and then he would pick him up.

The site Long selected looked good from above, but it turned out to be full of obstacles, not to mention still behind enemy lines. Long glided as far as possible and then attempted to jettison his canopy at about five thousand feet but was unsuccessful. "There was nothing to do but attempt a belly landing."

In the meantime, three unidentified, single-engine aircraft took off from a nearby field, and Daniels' attention turned to them. When they headed in a different direction, he looked back for Long "but had no luck. The last time I saw him, he was on his approach leg into what appeared to be a suitable field for a good belly landing" at about 12:30 p.m. No search could be undertaken in enemy territory, near Szombathley, Poland.

"Skimming over treetops and dodging through spaces between trees, I finally hit the ground and began skidding. From this time on, I remember nothing about my landing," Long told Charles Francis. The next thing he remembered was running from a group of angry civilians chasing him.

Long had his pistol with him and recalled how other POWs with weapons had been severely beaten or killed, so he threw it away. The mob overtook him and began beating him until German soldiers arrived and intervened. The soldiers searched him, found the extra clip of ammunition for his pistol and wanted to know where the weapon was. A child in the village served as interpreter, as the German soldiers attempted to question Long, but it didn't get far.

Long was taken to the town doctor, who bandaged his nose, which he suspected was split open by the cockpit gun sight when he crashed. Then a Hungarian colonel attempted an interrogation. Long knew the German language, so he understood most of what he was hearing. So, when a question came up that he didn't want to answer, he just pretended to not understand.

After five days in solitary confinement, he was treated for two weeks at a Budapest hospital. At the same time, the Allies began their campaign against the city, which angered the Nazis, and they put Long back in solitary. The next time he was interrogated, they had a book labeled "332nd Red Tails," practically a history of the group since its activation, Long said.

While in the jail, an inmate asked him why the Black soldiers were faithful to a country that treated them poorly, and Long replied with another question, "Why were the Germans so hard on the Jews?" The inmate's reply was that the Jewish Germans had sold out their country after the last war and had hoarded the nation's capital, insisting that Germany was the European underdog and had not been treated fairly.

After landing in Poland and being treated in Hungary, Long was taken to the German prison camp Stalag Luft III, a huge prison near Sagon, ninety miles southeast of Berlin. The camp housed primarily British and American airmen. Inside two layers of barbed wire fence were six compounds holding more than 10,000 POWs. Each barracks had one stove to 120 men.

Long was held in the center of what was called "north camp," primarily for British airmen and mostly bomber pilots. The Geneva Convention said that officer POWs were not to be forced to work, but they would volunteer to do things to make their life in the prison camp easier, like clearing tree stumps to make way for an athletic field.

These POWs did not fear for their lives, as long as they stayed inside the camp. The officers shared a library of more than one thousand volumes, and Red Cross parcels made life easier. Soap helped with "body lice and other creepy, crawly vermin," and other packages brought paper, pencils, games and sporting equipment. The Red Cross packages also augmented their meager German rations. "These were lifesavers, but the guys still dreamed of some good, old home cooking," fellow 332nd POW pilot Alexander Jefferson said.

In the POW camps, the German guards separated prisoners by officer or enlisted, not by race. "The Black pilots at the POW camps in Nazi Germany were in a more racially integrated environment than they would have been in most places in the American South," historian Daniel Haulman observed.

Many also dreamed of escape, but they were all too familiar with the fifty men executed for the attempted tunnel escape, immortalized in the movie *The Great Escape*. A loudspeaker and bulletin board reported announcements, such as arts presentations or athletic events, and a map showed the war's progress.

The Stalag Luft III camp was evacuated in January 1945, taking the prisoners about 360 miles to Stalag VII-A in Moosburg, Bavaria, where life was not as comfortable. With the Russians closing in, "The Luftwaffe officers in charge of our camp simply could not allow several thousand, highly-trained Allied airmen to be rescued," Jefferson said.

Although the temperatures during the march were as low as fifteen degrees below zero and they walked through up to eight inches of new-fallen snow, spirits among the airmen were high, particularly as they walked in new shoes, heavy socks, gloves, scarves and overcoats recently provided by the American Red Cross. After marching for days, they were packed into "40 and 8" box cars, arriving at Moosburg on February 3, 1945, along with thousands of men—Americans, British, Indians, Polish and Russians—from other prison camps, making for insufficient housing and poor sanitary conditions.

"In a sprawling set of tightly-spaced rows of drab, run-down, one-story military barracks built to accommodate 10,000 persons, the Germans had crowded together 110,000," according to POW Frank Murphy.

The filthy and miserable conditions were interrupted one day when a 332nd Red Tail P-51 flew over, while it shot up the Moosburg train station. From then, the prisoners could hear the nearing Allied forces. "Rumors kept us going, especially as the front lines closed in on us," Jefferson said.

After about an hour of intense firefight outside their fences, the prisoners were relieved by the sound of Sherman tanks crashing through the front gate on April 29, 1945. General George Patton himself arrived at the POW camp with a food truck and eighty dollars each in payroll for the American airmen.

Long had a fiancée in St. Louis, Margaret "Merle" Smith, to whom he wrote letters faithfully during his military training. Merle and her mother attended his graduation at Tuskegee. As best friends, Merle and Wilbur's sister Naomi kept in touch while Wilbur was overseas and held as a POW.

While in the POW camp, Long wore only the bloodstained clothing he arrived in and the scarf given to him by Merle. Trying to resume his life after returning to the States, Wilbur called Merle in St. Louis several times before she informed him that she was engaged to someone else. "He didn't show his feeling, but I sensed his tremendous disappointment," Naomi wrote. "I'm sure his thoughts of her and the promise that came with the engagement ring had sustained him through his roughest moments in prison camp."

During the many months of Wilbur's captivity, his family was comforted with a penciled letter from Wilbur at specified intervals. Their father had mysterious, but accurate, dreams, and their mother reread the 91st Psalm as they dealt with their son's uncertainty. Sister Naomi turned to her poetry.

"Suddenly war became very personal, and I became more painfully involved in it than had ever happened through news reports," Naomi wrote. The future poet laureate of Detroit, Naomi Long Madgett wrote the following poem in 1944 while he was "missing":

WHITE CROSS

Never mind of dawns refuse to waken you
Though the hills are sleepy blue with mist;
Never mind if lovers have forsaken you,
Seeming to forget the lips they kissed.
Heavy feet may thoughtlessly tramp over you,
Never caring that you had to die;
Silver wings of war may not discover you:
Many are the crosses where you lie.
But when Time is lying in the sod with you
And the stillness is its silenced drum,
There'll be one to seek the face of God with you—
Wait for me, my dear, and I will come.

Just before Naomi's college graduation in May 1945, the family learned that Wilbur Long had been liberated. At her dorm, the army's call, telling them that he was on his way home, was followed by a second from the airman himself saying that he was in the United States.

Long received an Air Medal Purple Heart. Upon his return, he served as public relations officer at Tuskegee Army Air Field and then as a P-39 pilot instructor at Selfridge Field, Michigan.

When fellow Red Tail captain Charles E. Francis interviewed Long, he was playing pool at Tuskegee Army Air Field. He was "small, slender, light complexioned [and] quiet."

"I was one of the fortunate pilots who got shot down and was lucky enough to get back home," Long told Francis.

Chapter 9

332ND FIGHTER GROUP

Wendell Pruitt, Richard Pullam, Finis Holt and Ozzie Farquhar

The pair of Wendell Pruitt and Richard Pullam had made it through pledging Lincoln's Alpha Phi Alpha, completing the Civilian Pilot Training and passing the U.S. Army Air Corps exams together. Although they enlisted separately, the two still wound up in the same class at Tuskegee Army Air Field in the spring of 1943.

After graduating together in December 1943, among the first one hundred Tuskegee Airmen fighter pilots, they were both assigned to the 332nd Fighter Group and stationed at Selfridge Army Air Field. However, they served with different fighter squadrons—Pullam in the 100th and Pruitt in the 302nd. Together they deployed with the 332nd Fighter Group in January 1944. While Pullam rose in leadership, Pruitt gained in glory.

The 332nd deployed from Selfridge Army Air Field only because of the success of the 99th Fighter Squadron disproving the doubts of white army air force leadership and Lieutenant Colonel Benjamin O. Davis Jr. advocating on their behalf with air corps command. Davis returned to the European theater in command of the 332nd Fighter Group, along with four other combat veterans from the 99th, who volunteered for another tour with the 332nd.

After a year of training and waiting at Selfridge, the 332nd left Hampton Roads, Virginia, in January 1944, aboard liberty ships, making the crossing to Italy in twenty-six days. During the crossing, Lieutenant Colonel Davis was the ranking officer aboard the ship, a new experience for most white officers and enlisted.

The 332nd Fighter Group first flew P-39 Airacobras when they arrived in Europe in February 1944. *National Museum of the U.S. Air Force.*

The 100th, 301st and 302nd fighter squadrons arrived in early February 1944 at Montecorvino Airfield, near Tarantalo, Italy, where they received hand-me-down P-39Q Airacobras from the 12th British Air Force. Before departing America, Pullam and his 100th squadron were able to visit the Bell Aircraft Plant, where the Airacobras were built. No surprise, then, the 100th was the first to begin flying.

Not long after the 332nd arrived in Italy, Pruitt and his wingman, Lee Archer, along with two others, allegedly performed some aerial acrobatics above Naples. Lieutenant Colonel Davis disapproved and ordered Captain Edward Gleed, commander of the 302nd, to court-martial the men. But "Gleed said he wouldn't willingly, because he thought it good practice and useful to combat pilots," recalled fellow Tuskegee Airman Charles Francis. Soon, Gleed was relieved of command and reassigned, but Pruitt and his wingman still had their wings.

Unlike the 99th, the 332nd encountered enemy attacks almost immediately upon arriving in the country. Many of their early missions were coastal patrol, convoy escort, harbor patrol, dive bombing and strafing in support of the ground troops in the area of Rome. The group was introduced to the realities of war early, losing its first pilot within the first month, due to poor flying weather.

Pullam arrived in Europe with four hundred flight hours, compared to the average fifty hours for newly arriving white pilots. "Those extra hours paid off. It meant pilots were so familiar with their P-39 fighter planes that the fuselage felt like another skin," Pullam said. "It was said the wings were attached to our bodies."

Capodichino to Ramitelli

In March 1944, the 332nd Fighter Group relocated to Capodichino Airfield, where the 99th Fighter Squadron was based. Pullam's 100th arrived first, along with Lincolnites in the ground crew—Ozzie Farquhar, mail clerk, and Finis Holt, radio operator.

It is likely that Lincolnites Bordeaux, Long, Pullam, Pruitt, Farquhar and Holt reminisced about their alma mater at some point before the 99th relocated the next month.

From Capodichino, these men would have witnessed the seven-day eruption of Mount Vesuvius beginning on March 22, 1944, which suspended flights for several days, caused airmen to wear gas masks and left a four-inch-deep layer of ash everywhere.

The 332nd Fighter Group learned that it would convert from the P-39 Airacobras to the P-47 Thunderbolts after a late April visit from the commanding general of the Mediterranean Allied Air Forces, Lieutenant General Ira Eaker. The visit also foreshadowed that their mission would change from strafing and patrol missions to bomber escort. The P-47 Thunderbolt "was more capable of engaging in aerial combat and had a range long enough to escort bombers on long missions," said historian Daniel Haulman.

The Capodichino field was raided on April 24, 1944, by nearly forty German Junker-88 fighters, wounding at least one ground crewman. Tipton native Jim Shipley recalled the night: "It looked like a bomb had been dropped on the air field; many of the planes were smoking and in flames."

Also in April 1944, Pruitt advanced to operations officer for the 302nd. Described as tall, handsome and Indian featured, Pruitt was "a youngster who was eager to become a pilot....His uncanny ability to maneuver the plane became so noticeable by other pilots of that group, that he became the envy of those who flew with him....He took chances that the boldest and most experienced pilots dared not take....Even professional acrobatic pilots were cautious in performing stunts that Pruitt did non-challantly," Francis wrote.

Pruitt would scrape rooftops with the wheels of his plane while flying at a high rate of speed and slow-rolled his plane so close to the deck that his wings missed the ground by only inches. "He seemed to have little or no premium on his life," Francis recalled.

Fellow pilot Charles Dryden called Pruitt "one of the top pilots of the Red Tails." Pruitt is featured on the History Channel's "Tuskegee Airmen" episode of *Dogfights*, and the character of Joe "Lightning" Little in the movie *Red Tails* may have been based on Pruitt.

CHESTER WHITE, AVIATION ENGINEER

In late May 1944, the 332nd group completed its transition to all P-47 Thunderbolts and was placed under the 15th Air Force. That required relocating all equipment and personnel to Ramitelli Air Field, on the east coast of Italy, closer to the bomber bases.

Ramitelli Air Field was just being completed by the aviation engineers when the 332nd began arriving. It had one airstrip, which was "basically just a dirt road that was covered with a type of steel mat," Jim Shipley said.

Aviation engineers were called on to build or improve an air field in advance of the squadrons and support crew arrival. Lincoln alumni Chester White may have built some of the air bases that the Tuskegee Airmen occupied.

White enlisted in February 1942 at Fort Des Moines, Iowa. By September, he had been commissioned as an officer from the engineer school at Fort Belvoir, Virginia. White deployed with the aviation engineers to North Africa in April 1943, remaining two and a half years. Serving under General Patton, he was part of the invasions of Italy and Southern France.

Illinois-born White wrote to the *Clarion* in April 1945 while serving in Europe, after reading about the bill in the Missouri General Assembly to abolish Lincoln University of Missouri. He said that he "spent [two] years overseas fighting against the very thing that, to me, Mr. Ellis' letter advocates. That is, I want to be recognized not as an American Negro citizen, but as an American citizen. Let's don't push ourselves under a racial bushel and hide our light."

Black aviation engineers built or improved more than one thousand air fields across the world. "Men who otherwise may have had little opportunity to work as heavy equipment operators or electricians, built airfields out of jungles, constructed roads over mountains and completed aircraft hangars on Pacific islands," U.S. Air Force historian Ronald Hartzer wrote.

The 332nd Fighter Group was stationed at Ramitelli Air Force Base in Italy in June 1944. *Toni Frissell photo, Library of Congress.*

Despite the huge potential following the war in this field, a board of officers in 1946 recommended that "because technical skills are relatively seldom attained by individuals of the colored race, Aviation Engineer units requiring a high proportion of technical skills would not normally be colored. On the other hand, colored personnel may be used, without comparable sacrifice of efficiency, in units wherein labor requirements are predominant."

After his discharge in June 1946, White went to work for the U.S. Army Corps of Engineers' St. Louis office, where he reached the rank of major as a program analyst.

FIRST BOMBER ESCORT

The 332nd Fighter Group flew its first mission of bomber escort on June 7, 1944. Two days later, Pruitt destroyed his first enemy plane, an ME-109, in a dogfight. His was one of the 332nd group's first five aerial victories, all that day in the Udine Area of northern Italy.

Lieutenant Colonel Benjamin Davis was awarded the Distinguished Flying Cross "for so skillfully handling his squadrons that only a few bombers were lost," that day, Haulman wrote.

The squadron had been flying the P-47 Thunderbolts for only a week and had yet to be formally trained by military and industry representatives. In the meantime, the squadron's own engineering officers and line chiefs figured it out and spread the word. The tardy training entourage arrived on the day of combat, while the 332nd was out on its second mission.

Staff Sergeant Samuel Jacobs, who was Pruitt's crew chief, said, "I remember this major standing atop a munitions carrier telling us 'boys' all about the 'flying bathtub' and how it should never be slow rolled below 1,000 feet, due to its excessive weight. No sooner had he finished his statement than 'A' flight was returning from its victorious mission. Down on the deck, props cutting grass, came Lieutenant Pruitt and his wingman Lee Archer, nearly touching wings. Lieutenant Pruitt pulled up into the prettiest victory roll you'd ever see, with Archer right in his pocket, as the major screamed: 'YOU CAN'T DO THAT!!!'"

On his third mission, June 11, 1944, Pruitt shot down his second ME-109 during a penetration escort to Munich, Germany:

> *As the Jerries passed under me, I rolled over, shoved everything forward, dove and closed in on one ME-109F at 475 mph. I gave him a short burst of machine gun fire, found I was giving him too much lead, so I waited as he shallowed out of a turn. Then, I gave two long 2-second bursts. I saw his left wing burst into flame. The plane exploded and went straight into the ground, but the pilot bailed out safely.*

To the 332nd Fighter Group, the white bombardment wing commander said, "Your formation flying and escort work is the best we have ever seen."

Throughout June 1944, the 332nd continued bomber escorts to Munich, Budapest, Bratislava, Bucharest and Sophia, plus strafing missions in Italy, Yugoslavia and Albania.

Pullam was promoted to a squadron leader the same month. During his overseas service, Pullam's only injury was a scratch to the forehead from a landing, where he lost air speed on approach and dropped twenty feet due to the full load of gas under his wings. "The tanks dropped down on the runway, gas flew up my face and I flew out of the cockpit, got on my feet and started running on the runway away from the plane," Pullam said.

SERGEANT FINIS HOLT

At Capodichino and Ramitelli air bases, each morning when the pilots left on missions, Finis Holt and his communications crew with the 100th Fighter Squadron climbed into their large truck filled with radio equipment and mounted with antennas. It was a silent job, listening in their headphones for word from the pilots, who rarely spoke in case the enemy was listening. Holt's equipment allowed him to keep in contact with pilots up to 750 miles.

Sometimes pilots would call in to ask about the weather; others would be disoriented after a mission and need guidance to make it home. Holt's truck did not return to base until the last plane, sometimes with white pilots, of each mission made it back.

Several nights, their base was raided by German bombers. When the air

raid sirens sounded, Holt and his fellow airmen jumped in their foxholes. He had never seen flares until the Germans dropped them from above. "It looked like the Fourth of July," Holt said. He recalled one raid that destroyed several planes on the ground and lasted forty-five minutes, according to the *Stars and Stripes* newspaper. "It felt to me it lasted two days," Holt said. The next morning, they discovered a shoebox-sized piece of shrapnel lying less than six feet away from their foxhole.

A New Bloomfield native, Finis Holt earned his bachelor's degree in economics from Lincoln in 1938. Expecting to work his way through graduate school, Holt found jobs in his field scarce at the tail end of the Depression, eventually finding work at the Albion (Michigan) Malleable Iron Company in early 1941.

New Bloomfield's Finis Holt graduated from Lincoln in 1938 with an economics degree. He served as a radioman for the Tuskegee Airmen's 100th Fighter Squadron in Europe, earning seven Bronze Stars. *Lincoln University Archives.*

Having registered for the draft in Chicago, Illinois, Holt returned there when he was called to active duty. He was inducted at Fort Custer, Battle Creek, Michigan, and completed basic training at Jefferson Barracks, Missouri. Holt chose radio school for his specialty. At Scott Air Field, Illinois, he learned to produce international codes messages at twenty-five words per minute.

At Tomah, Wisconsin, he was part of the first Black trainees to attend the school for advanced VHF equipment training. The experience was integrated until it came time for commencement, Holt recalled in an oral history interview with the Albion Library. The base tradition was to hold a dance. With no Black women nearby, the Black airmen were given five days' leave in Chicago instead of attending with their white comrades.

"It was a good experience, but war is still hell," Holt said.

UNBELIEVABLE FEAT

On June 25, 1944, twelve planes led by Captain Joseph Elsberry left on a mission to patrol roads in Northern Italy, expecting reinforcements to come from Yugoslavia, with the primary target being Ancona, Italy. The 332nd squadrons were to fly up the coast of Yugoslavia to the Pola area (in present-day Croatia). But strong winds over the Adriatic Sea blew them off course toward Trieste Harbor.

Their mission briefing assured the pilots that they would find no Allied vessels in that area. In the harbor, the flyers spotted what appeared to be a small American destroyer. When the ship—which turned out to be an *Ariete* class Italian destroyer—fired at the flyers, they discovered that it was not. A thick blanket of smoke developed over the 240-foot-long former destroyer *Giuseppe Missori*. The pilots attacked.

The P-47s dove in pairs, with Elsberry and Lieutenant Henry Scott taking the first run. Next went Lieutenant Joe Lewis and Lieutenant Charles Dunne. On the third wave, Pruitt made the first direct hit, setting the ship on fire. Pruitt's wingman that day, Lieutenant Gwynne Pierson, followed up with another direct hit, causing the magazine to explode.

"Dad's gun camera (I've seen the film) shows his tracers hitting the water, walking up the side of the boat, and then going in an open hatch—which he thought was strange since boats under attack were supposed to secure all openings. The rounds went in the hatch, and the ship blew up. He always said his first thought was that he had to get around/over the explosion," Pierson's son Scot posted on ww2aircraft.net.

"Up to this point in the war, no other fighter aircraft had ever [permanently disabled] a destroyer with machine gun fire," Scot Pierson said. In fact, many white leaders did not believe that the Black pilots had managed the feat until they saw the wing cameras confirm the success. Once confirmed, Pierson and Pruitt received the Distinguished Flying Cross.

The Iconic Mustangs

The last week of June 1944, the 100th, 301st and 302nd Fighter Squadrons were equipped with the more maneuverable P-51 Mustangs. With them, they earned the famous nickname of "Red Tails," when the plane tails were painted with the iconic solid red to denote them from the other flight groups in the 306th Wing of the 15th Air Force.

The 99th Fighter Squadron joined the 332nd on July 3, 1944, at Ramitelli. The 332nd Fighter Group now had the distinction of being the only fighter

Captain Wendell Pruitt would give his crew chief, Sergeant Samuel Jacobs, his ring before takeoff on each mission. Behind them is the Mustang *Alice Jo*, in which Pruitt shot down three enemy planes and helped destroy an Italian *Ariete*-class destroyer with machine gun fire. *Library of Congress.*

group with four squadrons. The Red Tails of the 332nd group were involved heavily in the Operation Anvil invasion of Southern France through July 1944, guarding B-24s and B-17s as they bombed strategic locations.

The flyers were observant. Pullam said that they noticed "the Germans would use a different burst of flak to signal the end of shooting and the imminent arrival of fighter planes....We got smart enough and when we saw that burst, we were ready to meet them."

By the end of July 1944, Pullam had been promoted captain, and Pruitt's promotion followed in August. Each often were mission flight leaders for their respective squadrons.

For their "heroic actions" on August 27, 1944, Pruitt and two other Black pilots were recognized "for exceptionally meritorious performance in the skies above bomb-scarred Europe." Pruitt and five other Mustang pilots under Colonel Davis destroyed twenty-two enemy aircraft on the ground at a Czechoslovakian air field on their return from an escort mission with B-24s to Blechhammer.

"Undoubtedly the most colorful and daring pilot of the group is Capt. Pruitt of St. Louis....Pruitt is a flier's flier, according to veteran airmen. A natural saying at the base is: 'If Pruitt can't do it, it can't be done.' Between missions, he flies acrobatics above home base for relaxation. The 23-year-old veteran is the idol of every young pilot in the group," the *Pittsburgh Courier* reported.

Pruitt also was a flight leader on August 30, 1944, when the 332nd destroyed eighty-three enemy planes and damaged another thirty-one.

OCTOBER 12, 1944

One of the most productive days for the 332nd was October 12, 1944. All four squadrons were involved in the mission, escorting bombers to Blechhammer, Germany, with both Pullam and Pruitt among the seventeen squadron leaders, including Colonel Davis himself. (One month earlier, fellow Lincolnite pilot Wilbur Long flying with the 99th Squadron was taken as a POW after reaching the same target.) This day, the 332nd had nine air kills, destroyed twenty-six enemy ground aircraft and damaged another sixteen at two German air fields. They also took out trains, trucks and oil barges.

Pruitt and his wingman, Lee Archer, were known as the "Gruesome Twosome." On this day, the pair accounted for five of the 332nd Fight

Group's nine air victories. Pruitt also was given credit for destroying six more enemy planes on the ground.

Of the fifteen-minute dogfight between two Red Tail Mustangs and a dozen German ME-109s, Archer later said that they "had no concerns about tackling a numerically superior force." In the initial formation, the 302nd was low and to the extreme right, Archer explained to Francis. Just after crossing Lake Balaton, Archer spotted enemy flyers "at 2 o'clock on the tree top." When he "called in the bandits," Pruitt was the first to respond.

Pruitt reeled off from formation, rolled his plane over and dove for the enemy aircraft, with Archer close behind. The St. Louis pilot made two passes at a Heinkel-111 bomber, and after his third pass, the enemy was smoking and Archer finished it off. The enemy formation then headed their way. "Instead of avoiding them, Pruitt flew directly into the formation with his guns blazing away," Archer said.

Once the pair made it through, they made a tight turn, positioning themselves behind three enemy planes—Archer took down one, and Pruitt's aim reached a second. Pruitt was chasing a third enemy plane when a German ME-109 got behind him. Archer protected his wingman and earned his second victory of the day.

When Pruitt's guns jammed in his pursuit of what would have been his third, Archer flew into position and scored his third of the day. However, debris from this last encounter damaged Archer's propeller. Pruitt escorted him to an island, where Archer landed and made repairs.

Archer recalled that he "almost met death trying to follow Pruitt" that day, despite arriving at the base feeling very good about their combined five victories. "They decided to do a little acrobatics coming in to the base," Francis said. It was customary within their group to do a slow roll over the field when one had a victory. This pair on this day buzzed the field twice before making their victory roll.

Pruitt, being left-handed, slow rolled to the right, unlike most pilots, while Archer slow rolled to the left. Archer's plane was upside down when it slid under Pruitt's. "Archer's prop stalled and, as he fell out of the roll, his plane's wing missed the ground by inches," Francis said. "I was through for that day and today consider myself one of the luckiest pilots in the world," Archer said.

Pullam also was decorated for his work on "Field Day," October 12, being awarded for meritorious achievement in aerial flight, while participating in sustained operational activities against the enemy.

CLARENCE "OZZIE" FARQUHAR

While pilots like Pullam and Pruitt rotated home after so many aerial combat missions, ground crew like Ozzie Farquhar remained in country. A monument apprentice before enlistment, Farquhar studied woodworking at a university in Florence, Italy, during his service. He was decorated with a Silver Star and three Bronze Battle Stars.

A fondly remembered mail clerk of the 100th, Corporal Farquhar returned home in mid-October 1945. After the war, Farquhar followed his high school sweetheart, Juel, to California. Farquhar "faced discrimination looking for work; and he labored at different jobs until he was finally hired as a mail carrier; a job he'd enjoyed while serving," with the Tuskegee Airmen, as his wife told the "Shouting from the Margins: Black Orange County, 1960 to 1979" project.

Clarence "Ozzie" Farquhar (left) was mailman for his unit in the 332nd Fighter Group in Europe and was the first Black postal carrier in Fullerton, California, after the war. *Kathy (Farquhar) Ayeh Collection.*

"Although he consistently scored the highest for the exam, postal officials hired the second and third white candidates to avoid hiring him as the government's rule required they pick from the top three candidates. Finally, a citizen, Suzanne Dean, intervened and wrote to the Postmaster…on his behalf. Clarence Ozel Farquhar was the first Black postal carrier in Fullerton. [His wife was the first Black teacher in Fullerton.] For many people, he was their introduction to Black human beings and served an invaluable role for his community," according to "Shouting from the Margins."

ROTATING HOME

Pruitt was one of the first pilots in the 332nd group to complete seventy missions, allowing him to rotate home. "He left a legacy in Italy—a record that was the envy of all and a list of accomplishments to be remembered by those who served with him and a goal for those who serve after him," according to the Pruitt chapter of American Veterans (AMVETS) website.

In late November 1944, Pruitt met with the mayor of St. Louis, Aloys P. Kaufmann, received congratulations and found that a day had been set aside in his honor. "He is typical of the fine young Negro men who have gone from St. Louis to war," the mayor said. "I'm going to set aside one day of your 21-day furlough when the city will honor you."

In his nearly three years of service, Pruitt, just twenty-four, flew seventy missions, received the Distinguished Flying Cross and was awarded the Air Medal with seven oak leaf clusters (dates for his distinguished service were June 30, August 6, August 22, September 7, two from November 1 and November 21). In 350 combat hours, he was credited with shooting down three German planes, destroying eight more on the ground and helping to permanently disable a Nazi destroyer.

"Modest about his overseas service, Capt. Pruitt chalked up his hits to 'lucky shots.'…he once had fooled the enemy by flying upside down…. Capt. Pruitt admitted he was proud of his part in the sinking of the German destroyer, a feat the mayor called the first-of-its-kind success [and] a fine piece of work," the *St. Louis Star Times* reported.

Pruitt went on to note that he had been fortunate to neither be sick nor injured during his nearly ten months overseas. "The going got pretty rough at times," he admitted. "Several bullet holes were shot in his ship, a P-51 Mustang, but he was not struck….His Mustang fighter plane is called the *Alice Jo*, for his fiancée, Miss Alice Charleton, Detroit, a civilian pilot, and for Josephine, the girlfriend of a mechanic who services his plane."

Pruitt was the first Black fighter pilot from St. Louis to return home on leave after combat. In response to Mayor Kaufmann's request for what the city could do for the air hero, Pruitt said, "I just want to rest. It sure is good to be back. When they told me to go home, I didn't want to do it, and they ordered me. But, now that I'm here, it sure is nice."

December 12, 1944, was designated as "Capt. Wendell O. Pruitt Day" in St. Louis City. James Cook, executive secretary of the Pine Street YMCA, was chairman of the arrangements committee. A big celebration was planned, with a noon parade through part of downtown, ending at a reviewing stand at Soldiers' Memorial. "Units from Scott Field, Fort Leonard Wood, the Civil Air Patrol, American Legion, cadet nurses and other organizations had been invited to participate" and schoolchildren were going to be excused to view the parade.

Instead, weather relocated the tribute to St. Louis City Hall. Hundreds filled the first-floor rotunda and promenades of the upper three floors to see the ceremony. Pruitt was presented a gold watch on behalf of the Black citizens by N.A. Sweets and a fountain pen from the pupils of Turner School.

Mayor Kaufmann said, "Capt. Pruitt 'typifies the patriotism and valor of the people of his race' and that he has won honor for himself and his city 'by his daring and skillful deeds.'" American Legion Department commander A.D. Welsh said that Pruitt "through his service record, had enlisted the aid of his fellow Negroes in the fight against anti-Negro measures and sentiment." In reply, Pruitt said, "The real heroes are the boys still over there."

Another dignitary of the event was Pruitt's flight instructor from his days in the Lincoln CPT program, John Randolph, who was manager of Lambert–St. Louis Airport at the time. Representing Lincoln at the St. Louis event were President Sherman Scruggs and Professor James Freeman, as well as a student honor guard including war veterans George Brooks, Alphonso Hilliard and Bernard Bates, plus students Preston Cathorn, Robert Brashear and Lawson Palmer.

Two days later, Pruitt was honored on the Lincoln University of Missouri campus on December 14, 1944, with a special convocation in Page Auditorium, sponsored by the Committee on War Information, chaired by Dr. Lorenzo Green. The women's league and the men's club held a tea and smoker for him, respectively. And then Pruitt and his mother were guests at a Christmas party at the home of President Scruggs.

Pullam Returns Home

Pullam returned to the States just a few weeks after Pruitt, having received the Air Medal with four oak leaf clusters and two major campaign stars. He was proud of the 100th, which flew 143 missions protecting American bombers, Pullam said.

His happiest memory, however, was "when they told us we were all coming home," Pullam said. Unlike Pruitt's fanfare, Pullam was met quietly at the Kansas City airport by his son A.E. Pullam, born in March 1944, and his wife, Martha Helen Cole, who had been staying with her family in Columbus, Mississippi.

While on leave in Kansas City, Pullam spoke to many groups, attended special events in his honor and toured defense plants and schools. In January 1945, he was a guest at Lincoln's 79th Founder's Day Convocation, where he made some brief remarks and was presented with a war bond.

After their well-earned furloughs, both Pruitt and Pullam took up their new posts as flight instructors at Tuskegee. Both captains and veteran pilots of Mediterranean operations, they completed a four-week refresher course before resuming duties in early April 1945.

Startling Tragedy

Soon afterward, a tragedy was reported in the April 16, 1945 *St. Louis Star and Times*: "Capt. Wendell O. Pruitt, Hero of Air War, Is Killed in South."

Pruitt was piloting an advanced trainer plane with Private Edward N. Thompson, an airplane mechanic, on board. At about 3:30 p.m., Pruitt made a normal landing. Thirty minutes later, the AT-6 took off again, making a steep climbing turn.

At 4:15 p.m., Pruitt buzzed Kennedy Field, coming down to an altitude of about ten feet over the field. He pulled the nose up and started a slow roll. The ship was at an inverted position at approximately seventy-five feet altitude when the motor cut out and the nose of the ship dropped. The ship crashed in an inverted position about eight hundred yards southeast of the field, where it burst into flames.

"Although most of his comrades had witnessed many pilots meet the same fate, they could hardly believe that this had happened to Pruitt," Francis said. The explanation among the fellow pilots was that "Private Edward Thompson became nervous and froze the stick when Pruitt attempted a slow roll on deck."

Wendell Pruitt looks out from his cockpit. *Library of Congress.*

While home in December, Pruitt's family had asked him in what circumstances would a pilot abandon his ship. His answer was: "A good pilot never abandons his ship. He always feels there is something to do to bring it to safety."

It was left to Pullam to accompany Pruitt's body from Tuskegee to St. Louis. Services were held at St. Elizabeth Catholic Church, and he was interred at St. Peter Cemetery, Normandy. Joining Pullam as a pallbearer was fellow Lincoln CPT cadet Lieutenant James Garrett.

Several hundred people attended Pruitt's funeral, where Father Andrews of St. Elizabeth Catholic Church eulogized Pruitt's "good qualities, keen intellect…his daring sense of honor, energy, citizenship, friendship and religious faith."

As noted in Francis' *Tuskegee Airmen*, Andrews continued, "Pruitt is dead, but the fruits of his life will be multiplied over the earth. He was a student, a man who used his intellectual abilities to achieve. All of us could learn from Wendell Pruitt the necessity of labor to accomplish our goal. We knew him as an interesting and unassuming young man, daring but not bold, vigorous and energetic, but not offensive, a good citizen, a good friend and a devout Catholic. We needed a modern hero for us to pattern our lives, so God called him home."

Chapter 10

477TH BOMBARDMENT GROUP

Stewart Fulbright, Victor Barker and Albert Gaines

Growing up in Springfield, twelve-year-old Stewart Fulbright Jr. walked three miles to the local air field when he learned of an air show. Aviation was so new that the air field didn't have runways yet. The flyers asked the gathering children for volunteers to go up with them. "I was fascinated with the idea I could get into an airplane," Fulbright told Fox News.

When it was Fulbright's turn, he sat in the front seat of a biplane during a five-mile pursuit race. The door did not shut completely, allowing him to see the ground below him. He had no parachute, yet he had the gumption to stand up from his metal seat to look around. A very anxious pilot gestured for him to sit back down. "I wish I could have stood the entire trip!" Fulbright said.

Knowing that his parents would not have approved, Fulbright did not tell them about his first flight until after he had graduated from Tuskegee Army Air Field.

Like many aspiring pilots in the 1930s, Stewart Fulbright read *G8* and *Flying Aces* and built his own model airplanes. "Growing up, flying was always in my mind.…I wanted to fly but never thought I would have an opportunity," he told Lincoln's KJLU radio. Although he was a student at Lincoln while the CPT program was offered, Fulbright said, "I could never get the amount of time or money together that it took to get into the Lincoln CPT program."

When the U.S. Army Air Corps was opened to Black airmen, Fulbright immediately volunteered for two reasons. The first, obviously, was to chase his childhood dream. The second was due to the waiting list, which meant he could stay in Jefferson City longer.

Not only did he reach the rank of captain, but Fulbright was also in charge of a six-man crew on a B-25 Mitchell medium bomber with the 477th Bombardment Group. "It turned out to be a lot of fun," he said.

Fulbright's story is part of the Library of Congress' Serving: Our Voices oral history project. He was among fifteen African American veterans to be featured in its "Equality of Treatment and Opportunity: Executive Order 9981" collection. President Harry Truman signed the mandate of "equality of treatment and opportunity for all persons in the armed services without regard to race, color, religion or national origin" on July 26, 1948.

Before that, however, in the U.S. Army Air Corps, equality was growing with each post exchange (PX), cafeteria or officers club, where whites-only policies were crushed by Black airmen refusing to accept the pattern of segregation.

The lesser-known part of the Tuskegee Airmen legend, the 477th Bombardment Group remained stateside but helped move many racially motivated roadblocks in the U.S. Army Air Corps. For example, instead of setting up all the necessary specialty schools for Black airmen separately, roles like bombardiers, navigators and medical personnel trained at previously whites-only bases and continued to challenge segregated services and spaces on those bases.

"They didn't believe that young, Black men had the ability or the courage to learn how to fly and to fight, and so all kinds of barriers were set up to keep it from succeeding," Fulbright told KJLU. He used the experience of racism in the military as an "ego-builder. We took it as if we could get through a program as difficult as that was, that we could do anything." Fulbright grew from doubting his own abilities to believing that he could be successful at whatever he tried in the future…except the hula hoop, he joked.

Fulbright earned his wings in December 1943 with class TE-43K (twin engine, 1943, November), as part of the second class to train in twin engines at Tuskegee Army Air Field. Only thirty-one from his original class of fifty-five graduated. "It was pretty rigorous. It was hard to get in, hard to get through," he said.

After graduation, Fulbright was assigned to Mather Field, east of Sacramento, California, for transition training to the B-25 Mitchell medium bombers, introduced in 1941 for Allied air forces and notably used for the April 1942 Doolittle Raid on Japan. He was among the first fifteen Black pilots at Mather Field and its only integrated class.

The Springfield-born Fulbright Jr. was Lincoln's first and only pilot in the 477th.

Specialty Fields

The 477th was designed to have four bomber squadrons, with 1,200 officers and enlisted men operating sixty B-25 Mitchells. Its permanent base was Selfridge Field, Michigan, where airmen arrived after completing their skilled training at sites across the nation. The bomber group occupied the same base as the 553rd Fighter Squadron, which held replacement P-39 pilots for the 332nd Fighter Group after the latter deployed overseas.

"Most of us would like this twin-engine training, especially for its postwar benefit, but there are quite a few of us who still like the single engine fighters," fellow Lincolnite Louis Harris wrote to the *Clarion* from Tuskegee in November 1943.

The B-25 Mitchell required a pilot and crew of five—co-pilot, bombardier, navigator, radioman and engineer. Many of the onboard crewmen began their military career in the flight training program at

The 477th Bomber Group flew B-25 Mitchells, like the Experimental Aircraft Association's *Berlin Express*, which visited Jefferson City, Missouri, in 2023. *Amy C. Nickless.*

Tuskegee, like Lincoln CPT cadet Victor Barker. The 477th Bombardment Group expanded the opportunities for these trailblazers to continue the specialized training and earn officers commissions.

Navigators trained eighteen weeks at Hondo Field, and bombardiers trained at Midland Field, both in Texas. Gunners trained at Eglin Field, Florida. Other Black bomber crew for the 477th were trained at bases near Roswell, New Mexico; Sioux Falls, South Dakota; Lincoln, Nebraska; and Scott Field, Illinois.

VICTOR BARKER

A trombonist with the Collegians while earning his math degree at Lincoln, Victor Barker paid for his education by teaching and writing music. He was born in Kansas City, the son of a minister, and was a major in his high school ROTC.

Barker taught at the Lincoln Lab High School for the 1940–41 school year and then worked in electronics and radio mechanics in Kansas City. He was training at the West Virginia Institute as a radio repairman before enlisting in the Signal Corps in August 1942. By mid-1943, he was at Tuskegee Army Air Field.

Victor Barker earned ratings as a navigator, bombardier and aerial gunner with the 477th Bomber Group and retired as a flight officer. He was inducted into the Lincoln University Alumni Association Hall of Fame. *Lincoln University Archives.*

In early 1944, he was "trained in the development of radar, mobile communications and deciphering machines," while at bombardier training at Midland (Texas) Army Air Field. Later, he added the rating of navigator from Hondo (Texas) Army Air Field.

After his navigator and bombardier days were over, Victor Barker became a math teacher in Kansas City and later a chemistry professor before retiring from the Department of Agriculture. He was a founding member of the Heart of America Chapter of the Tuskegee Airmen International and active in the Lincoln Alumni Association. In 2012, Barker was posthumously inducted into the Lincoln University Alumni Association Hall of Fame.

Albert Gaines

Another Lincolnite, Albert Gaines, also earned his wings as both navigator and bombardier, but in the reverse order as Barker. Gaines joined Barker at Midland Field, where the former earned his bombardier's silver wings in January 1945.

Gaines was born in Ohio, where his father worked in a steel hoop factory. He arrived at Lincoln in the fall of 1941, and the next year he was working in a factory in New York. He enlisted in August 1943 and was training at Hondo Army Air Field by April 1944. Gaines took first place in an endurance marathon with twenty-three other navigation cadets wearing full field infantry packs and carrying rifles. In August 1944, Gaines earned his silver navigator's wings with only the second class at Hondo Field to include Black cadets. Gaines went on to reenlist in the new U.S. Air Force Reserve in September 1947 for three years, progressing to the rank of captain.

Joining Barker at Midland Army Air Field, Gaines graduated advanced bombardier school in January 1945. In a letter to the editor of the *Clarion*, he wrote in December 1944 from Midland, Texas, "My stay at Lincoln was short and it holds the most pleasant memories of my life....I see in your articles the same spirit and determination that has made Lincoln what it is....I'm hoping that [Corporal Russell Estill Jr.] and I will soon be able to visit our beloved Lincoln....At every base where I've been stationed, I've run into Lincoln women and men and they are doing a grand job."

Of Barker and Gaines at Midland Army Air Field, Estill extolled that they "are leaders in scholastic averages for their respective classes. This exemplifies the background and training which one receives at Lincoln and which causes that institution to be a beacon light for the rest of the colleges, as far as scholastic competition is concerned."

Born in Fayette, Estill graduated from Lincoln University High School in 1938 and then attended Lincoln University. Before being assigned as a noncommissioned officer at Midland, Estill was stationed in Winfield, Kansas, with the 322nd Aviation Squadron, where he was active with the entertainment committee arranging dances, jive sessions, programs and out-of-town talent, musical and dramatic events. He was a member of the squadron's quiz team and basketball squad before moving to Midland Army Air Field in November 1944. Estill left the service as a tech sergeant and later became a founder and president of the Detroit Chapter of the Lincoln Alumni Association.

THE FIGHT AT HOME

While the 332nd Fighter Group was fighting enemies in Europe, members of the 477th Bombardment Group and other Black airmen still stateside fought for equality at home. "We were faced with intense segregation while in training down at Tuskegee and always after it," Fulbright said.

Every U.S. Army Air Corps field across the nation had been whites only until World War II. Despite air corps leadership issuing orders against segregation, most base commanders found ways to perpetuate prejudice. Meals and housing were almost always segregated. Training was the first aspect to be integrated.

The off-duty facilities were the catalyst for most Black protest for change on a given base. For example, at Hondo Army Air Field, the Cadet Club only served white cadets inside, forcing Black cadets of the same rank and training status to order food at the back door. At Midland, Black officers were prohibited from using the officers club while it was used for a dance where white women would attend.

Black cadets at Hondo Field often chose to travel to nearby San Antonio, Texas, for entertainment and religious services with the Black community and Black USO. But, air fields in Yuma, Arizona, or Roswell, New Mexico, did not allow such freedom.

SELFRIDGE ARMY AIR FIELD, MICHIGAN

Just before the 477th's activation in January 1944 at Selfridge Field, Black officers from the 553rd made repeated attempts to integrate the base officers club without violence. Pilot Everett Bratcher and Tipton native Jim Shipley were witness to one of the many decisions by white command to prohibit equality for Black officers in the U.S. Army Air Corps.

Major General Frank Hunter issued an order prohibiting Black officers from using officers clubs at Selfridge, which was in violation of the orders already issued at higher command that all air corps officers were to be treated equally. A sit-in was organized by nearly one hundred Black officers, who were arrested and threatened with court-martial, but the charges were dropped.

That led to the appointment of Lieutenant Colonel Benjamin O. Davis Jr. as commander of the 332nd Fighter Group, returning from his successful leadership of the 99th Fighter Squadron in Europe and his triumphant lobbying of military leadership to combat racist attempts to block Black pilots from combat duty.

Two new service clubs were planned, but never implemented, in March 1944—a Black officers club and one for Black enlisted men. At the same time, Brigadier General Benjamin O. Davis Sr., the first Black general in the U.S. Army, visited the base to investigate the treatment of Black personnel. A few days later, the commander who had refused Black officers to enter the "white" officers club was replaced.

Major General Frank Hunter, commander of the 1st Air Force, visited Selfridge Field in April to make his own assessment. Rather than lay blame with the base commander for going against U.S. Air Corps regulations, Hunter ordered a new club be built for Black officers. However, the 477th and 553rd were transferred before it was completed. The war department eventually reprimanded Colonel William Boyd, Selfridge Field commander, for failure to enforce air corps regulations allowing all officers to use the same club.

All of the major Black units stationed at Selfridge Field were relocated in May 1944. The 477th Bombardment Group—including Fulbright, Barker and Gaines—was transferred to Godman Army Air Field, Kentucky. The 553rd Fighter Squadron, whose weather officer at the time was Lincoln's Charles Anderson, was sent to Walterboro, South Carolina, and then disbanded.

Freeman Army Air Field, Indiana

The relocation did not resolve the constant issue of racial inequality in the military. Godman Army Air Field, Kentucky, was a smaller base. The white officers from Godman chose to travel a few miles to the officers club on the neighboring Fort Knox base, leaving the smaller Godman field officers club for the segregated use of Black officers. The 553rd's experience in Walterboro, South Carolina, was similar, as white officers rented a club for themselves off base.

Godman Field was highly inadequate for the needs of a bombardment group preparing for overseas duty. Nevertheless, the airmen limped along there for ten months. By March 1945, the 477th Bombardment Group reached full strength, and plans suggested that it would deploy to the South Pacific that summer.

They were relocated to Freeman Army Air Field in Indiana, where 117 acres and four large runways made a better fit for military training. But, the racial insults continued with the return of separate facilities for "trainees" and "supervisors."

This attempt at segregation, despite the U.S. Army Air Corps' explicit desegregation orders, defied logic. Several pilots of the 477th were veterans of the European theater. Clearly, combat veterans with the rank of first lieutenant or captain were not "trainees." Likewise, not all white officers were "instructors," yet white officers fresh from Officer Candidate School were given the privileges of instructors. The three combat veterans of the 99th pursuit squadron who had arrived at Freeman Field in March with the 477th were quickly and quietly assigned off-base duty.

Fulbright described the "trainees" club at Freeman Field as a little black building with a heating stove in the center and three card tables. By contrast, the "base and supervisory" personnel gathered in a large, white brick officers club with steam heat and the traditional fittings of an army officers club. "We decided that this was not right, and that we were going to just go to the so-called white officer's club," Fulbright told KJLU.

Black officers of the 477th planned a nonviolent entrance to desegregate the white officers club. A few Black officers at a time attempted entry, thirty-six the first night and twenty-five the next. When they were turned away, another group took their turn. However, someone notified the base command of the plan, and military police were stationed at the club entrance. "By the time we got there, the club had been closed," Fulbright remembered.

These sixty-one men were arrested for disobeying an order from a superior officer. The next day, Colonel Robert Selway, commander of the 477th, ordered the officers club for "trainees" closed. The detained Black officers were released, except for three who were also accused of using violence. These three were court-martialed in July 1945; only Lieutenant Bill Terry was convicted for "jostling" a fellow officer while trying to walk past the guard at the club entrance.

Colonel Selway then issued a new base regulation that specified separate buildings to be used not only as officers clubs but also recreational buildings, mess halls, officers' quarters and latrines. This order also specifically defined the separation as "personnel undergoing training" (Black) and "base supervisory, instructor and command personnel" (white).

Every officer on the base, Black and white, was called into the commander's office to sign a copy stating that they understood the regulation. "Most of us at least...refused to sign it because we said we could read it, but we could not understand it," Fulbright said. In the end,

101 Black officers were arrested for disobeying a direct order when they did not sign it. Many of the other 300-plus Black officers wrote their disagreement of the order along with their signature.

In wartime, disobeying a direct order could have a penalty up to execution. The 101 were taken to Godman Field and placed under arrest while the War Department investigated. "Those of us who refused to sign it and who were not arrested were primarily first pilots and the reason we weren't arrested is because the commanding officer wanted to keep the reputation of having so many flight hours placed on the airplanes intact," Fulbright said. "However, most of us refused to fly....I, during that period, had a lot of dental work done, others had all kinds of excuses, we had to have some kind of seemingly legitimate reason for not flying."

The War Department, whose orders Colonel Selway had ignored by issuing his base regulations, released the Black officers one week after the signing boycott. However, Colonel Selway issued each an administrative reprimand upon their release. Those reprimands were removed from their personnel files fifty years later.

Godman Army Air Field, Kentucky

After only one month at the larger Freeman Field, anticipating deployment overseas, the 477th bomber group was returned to the smaller Godman Field in Kentucky. At the same time, in June 1945, Colonel Benjamin Davis Jr. was recalled from leading the 332nd Fighter Group in Europe to take command of the 477th Composite Group, which included fighter squadrons as well as the bombers. All of the remaining white command officers, including Selway, were transferred. That allowed Black officers to advance into the lower command positions of the squadrons and the base, which previously had been a revolving door for inexperienced, and often unqualified, white officers.

Colonel Davis also was made commander of Godman Field, becoming the first Black officer to command a major Army Air Forces base. The 332nd Fighter Group returned to this base from Italy soon after.

The 477th Bombardment Group was ready with sixty-four six-man crews—pilot, copilot, navigator-bombardier, engineer-gunner, radio-gunner and tail gunner—to deploy to the Pacific theater. The airmen were sent home on furlough in August 1945 with the expectation that they would deploy upon their return. In the meantime, the atomic bombs were dropped on Hiroshima and Nagasaki, Japan.

Despite all of the internal adversity, most men of the 477th, like Fulbright, "wanted to demonstrate how good we were," he said. When they learned about the possibility of deploying to the South Pacific, Fulbright said that he felt good about it, having trained for two years, waiting. Then, the war was over.

POSTWAR

Fulbright's best friend, with whom he graduated at Tuskegee, died in a training exercise not long before Fulbright left the service. "We used to do crazy things with a bomber," he said, such as flying under bridges. But after his friend's plane crashed in the Ohio River, Fulbright never enjoyed flying as much as he had. "Take off and landing are the most dangerous parts of flying," Fulbright said. "It becomes routine after you do it so many times... but you must always remain alert."

In October 1945, the 332nd Fighter Group went inactive, although the 99th Fighter Squadron remained active with the 477th Composite Group at Godman. Fulbright remained in the air force reserve through 1968, reaching the rank of captain. "Discipline, leadership, and the ability to follow leadership are those attributes which you of the Hiking 69th have an opportunity to develop and which are expected of you as soldiers of the U.S. army," Fulbright told the male freshman and sophomore students at the Lincoln Lab High School in November 1945, after his discharge.

"White-washed as a freshman, embarrassed as a sophomore and soundly defeated both as a junior and a senior have led me to believe that there are truly horizons unlimited for everybody," Fulbright said in his address a decade later at his alma mater's annual Men's Day convocation.

Fulbright was inducted into the Springfield Public Schools Hall of Fame in 2012. His son, Ed, said, "My father was a very modest man and believed in doing your best and letting the results speak for themselves."

Chapter 11

OTHERS IN SERVICE

Most cadets from the Civil Pilot Training program at Lincoln University of Missouri went on to serve in the military, if not as Tuskegee Airmen.

George E. Banks was born in New Frankport and was living in Saline County when he enlisted in August 1942. He deployed to Europe in December 1944 as a duty soldier and was awarded a Bronze Star for the Rhineland Campaign.

Lucius Banks was born in Mississippi and reared in New Madrid. He deployed in March 1943 with the 481st Port Battalion as a carpenter to the South Pacific. He was awarded the Bronze Star for the Campaign of Luzon.

Frank Bruce joined the U.S. Navy.

Robert Buck was a high school teacher before serving as an auto mechanic in the U.S. Army 6485th Quartermasters. He served three years in the European theater, receiving two Bronze Stars for campaigns in Northern France and the Rhineland.

James C. Butler enlisted in October 1941 at Jefferson Barracks and served as a radar crewman with the 234th AAA Gun Battalion until November 1945. He received a Bronze Star for the India-Burma Campaign. He deployed overseas twice, first to the Asian theater from April 1942 to December 1943 and then to the China-Burma-India theater from December 1944 to November 1945.

George Carter, born in Joplin, was a truck driver in the U.S. Army Corps of Engineers, 376th Battalion. He served two tours from July 1942 to October 1945. He received four Bronze Stars for campaigns in Southern France, Rhineland, Naples-Foggia and Rome-Arne.

Chester B. Cathorn was a tech sergeant in the U.S. Army, 13th Infantry, headquarters company. He enlisted in March 1943 from Harrisonville and served as a telephone switchboard operator. He received two Bronze Stars for the Rhineland and Central Europe campaigns, from February to August 1945.

Lester Davis joined the U.S. Army Air Corps.

Joe Gayles was a sergeant in the U.S. Army Air Corps military police and band.

William Graham was a warrant officer.

John Kincaide passed the U.S. Army Air Corps exams in early 1941, along with Pullam, Pruitt and Wallace. He left the U.S. Army as a sergeant, having worked at Barksdale Field, Louisiana. He is buried at Jefferson Barracks National Cemetery.

John People was a sergeant with the 213th Port Company, receiving three Bronze Stars for campaigns in the Rhineland, Ardennes and Central Europe. He is buried at Jefferson Barracks National Cemetery.

John Perry joined the U.S. Army from California; Henry Scarlett from Erie, New York; and Lance Barber from Oklahoma. James Garrett became an officer in the U.S. Army.

Herman Plummer was released from service as a staff sergeant in the U.S. Army Air Corps. After the war, he was branch president of the NAACP in Salem, Oregon, and was nominated to the Oregon legislature in 1953. He is buried at Willamette National Cemetery.

William Spencer served with the U.S. Army warrant officers.

Compton Taylor served with the 4253rd Truck Company and left the service as a first lieutenant in command of an antiaircraft automatic weapons unit. He was awarded one Bronze Star for the campaign of Northern France. He is buried at Riverside (California) National Cemetery.

Khamalaw White was a staff sergeant in France.

OTHER LINCOLNITES WHO DID not take the Civilian Pilot Training also joined the U.S. Army Air Corps.

Herman Stevens, Maurice Bellamy and Maurice Copeland served with the 46th Aviation Squadron at Scott Field, Illinois, in May 1943, and Bellamy went on to serve later in England.

Wilbert Bartlett was a sergeant in the U.S. Army Air Forces, 837th Engineer Battalion. As a construction foreman, he served in Europe from October 1945 to April 1946.

Robert Enoex left the U.S. Army Air Corps as a master sergeant.

William Herndon and Harold Sylvester Green served with the U.S. Army Air Corps.

Byron Johnson was a warrant officer with the 53rd aviation squad at Mather Field, California.

John R. Williams also served in the U.S. Army Air Forces.

Chapter 12

POSTWAR

With the surrender of Japan on September 2, 1945, the war was over. For many of the Black airmen, they hoped that a career in aviation was in their future. For all Black service members, especially those who experienced equality while overseas, they hoped to return to a grateful and less prejudicial home.

The Double Victory posed in 1942 by the *Pittsburgh Courier*—Victory Abroad and Victory at Home—was far from being reached. But there were some steps toward a more just United States being made.

Many of the students who left Lincoln in the midst of their studies to serve their country returned in the fall of 1946. They were joined by hundreds of newly enrolled veterans, taking advantage of the June 1944 Servicemen's Readjustment Act, or the "GI Bill," providing funds for World War II veterans for education, unemployment and housing. So many veterans tried to enroll that semester that Lincoln had to turn away about one hundred due to inadequate housing.

Such an influx of Black soldiers seeking the benefits of the GI Bill in the year after the war caused Lincoln to build new veterans housing, which CPT cadet Elmore Nelson helped with before earning his education degree.

The first of his family to exceed a sixth-grade education, Nelson attended elementary school in Hayti and Washington High School in Caruthersville. While at Lincoln, Nelson qualified for the U.S. Army Air Corps enlisted reserves in November 1942 and then was called to service with the 743rd Military Police (Aviation) Battalion before a space opened at Tuskegee.

Nelson was wounded when the battalion first arrived in northern Africa and then served until released in December 1945 at the rank of corporal. Back on campus, Nelson was president of the Rural Life Club.

After graduation, Nelson returned to Hayti, where he taught industrial arts and math at Central High School for seven years. Then he was principal for the next eleven, until taking a job with the Missouri Department of Education's Vocational Rehabilitation Division. In 1963, he served as the national president of the Lincoln Alumni Association.

As a student in the mechanical arts program, Nelson helped build eight housing units for 180 veteran-students, funded in part by the Federal Public Housing Administration. Temporary housing for 120 veterans was provided by moving Depression-era Civilian Conservation Corps (CCC) barracks from Fort Leonard Wood. But the new construction was not completed until the spring of 1947 due to a labor shortage. These veterans housing units were built in front of Damel Hall and the power plant, near Weir's Creek and the athletic field and near the old tennis courts. They had men living three to a room with a study area for every four rooms.

The federal government only paid for the erection of the buildings. That left the university to cover the cost of engineering and supervision, as well as most of the furnishings. Interestingly, the government funding provided ash trays but not writing tables, which the university's mechanical arts program built.

The Lincoln veterans housing units in the fall of 1948 were named "Jasonville," in honor of Lincoln's longtime vice-president. Each unit elected a "councilman," and all elected a "mayor."

Oklahoma City native James Tillman was mayor of Sycamore, one of the four units on Scruggs Avenue. Tillman served with the U.S. Army Air Corps, 462nd Aviation Squadron, in the Asiatic-Pacific theater, where he earned a Bronze Star. A gifted clarinetist, Tillman graduated with a music degree in 1956.

Among the GI Bill freshmen was Tuskegee Airman James T. Morris, who served with the 477th Bombardment Group. Morris was active in the Lincoln campus veterans organization, student council and Stagecrafters. He also played guard on the football team before graduating in 1949. Morris then served in the U.S. Navy from 1954 to 1969, including tours in Vietnam. He was aboard the USS *Ticonderoga* in August 1964 when it launched planes against four North Korean bases, for which the crew received the Navy Unit Commendation.

Former CPT cadets James Butler and Earl Hogan were among nineteen students to take flying courses at the Jefferson City Airport in the fall of

1946, offered by the Brummett Aircraft Company. The postwar, fifty-one-hour pilot training courses began in the spring of 1946 in cooperation with G.R. Cotton, the mechanical arts department and J.R. Brummet, manager of the Jefferson City Flying Field. "It will, as far as is known, be the first of its kind at an accredited college anywhere in the country," Brummet said in October 1945.

At least three other CPT cadets returned to Lincoln after their service. Alphonse Ellis served in the 26th Chemical Decontamination Company at Camp Carson, Colorado, and then completed his building engineering degree in the spring of 1947. Tuskegee Airman and CPT alum Lewyn Boler returned to Lincoln for a history degree in 1951.

Khamalaw White, who served as a staff sergeant in France, received a backdated football award in the spring of 1947 for his "sensational" work as halfback in the 1942 season. White played one more season of football in the fall of 1946, with fifteen other returning veterans under returning veteran Coach Hoard. After earning his degree in physical education, White coached twenty years at Inkster High School, Detroit, Michigan.

PROFESSOR CHARLES HOARD

Professor Charles Hoard only taught and coached at Lincoln one year before entering the U.S. Army Air Corps. Hoard returned after instructing physical education for Tuskegee Airmen at the historic air base in Alabama. He remained in the U.S. Air Force Reserve Squadron based in Jefferson City, where he reached the rank of major and was the squadron's training officer.

By the fall of 1946, he was back at the Lincoln Field, coaching track and football. Hoard also was the boxing coach, which meant that he was present at the Golden Gloves fight in Moberly on January 27, 1947, when student Charles Byas died. The match was called after one minute, seven seconds into the third round. The cause of death was a cerebral hemorrhage, the result of a series of blows to the head, the coroner said. Hoard confirmed that Byas had been pronounced physically fit by the school and that he had passed a preflight physical too.

Before the war, Hoard had been a men's hall proctor. When he returned, he became the dean of men and assistant physical education professor. He remained at Lincoln for his career, becoming a respected authoritarian and Christian model. After earning his PhD in guidance and counseling, he became dean of students and director of health service. He loved teaching, and in the 1960s, he became head of the department for psychology and education so he could have more time in the classroom. He sponsored activities and organizations on campus with a religious message.

"He greeted many students at the orientation by telling them to watch what they did while attending Lincoln University, or they would be sent home with an apple and a comic book. He was also known for stopping students as they walked throughout the community, reminding them that they must always act appropriately so as to not bring any shame to the university," said files in the Lincoln University of Missouri archive.

Charles Hoard returned to Lincoln in 1946, coaching track, football and boxing. He earned his doctoral degree in guidance and counseling in 1952 and then became dean of students and director of health services. Later, he was a professor of psychology and head of the education department. *Lincoln University Archives.*

Hoard met his future wife, Floretta "Yvonne" Walker, who was also a coach and physical education instructor, at Lincoln. They built in 1955 a brick ranch home, which is now a Historic City Landmark, at the northwest corner of Dunklin and Chestnut Streets. In the community, he was an original member of the city's Human Rights Commission in 1964. He was the first Black resident appointed to a public school committee after local NAACP leader Charles "Lefty" Robinson protested the whites-only appointments by the school board.

The university dedicated Charles Mason Hoard Residence Hall in 2001 on East Atchison Court.

REMEMBERING THE FALLEN

On the first anniversary of Armistice Day, the Lincoln campus' American Veterans Committee, representing 412 veterans enrolled at the school, held a dedication on the quadrangle to recognize the four Lincolnites fallen in service, with a plaque from the office of public relations presented by President Scruggs to committee chairman Walter Pinson.

Just before the Pearl Harbor attack, Richard Parker, the first of five Lincolnites to be killed in action, set aside his education to join the U.S.

Navy. Soon after that, he was stationed at Pearl Harbor, working as a mess attendant. Parker was aboard the USS *Lexington* on May 8, 1942, at the Battle of the Coral Sea. The American crew had taken down 40 enemy planes before 108 Japanese planes attacked the aircraft carrier. Two enemy torpedoes caused the thirty-three-thousand-ton vessel to be rocked with internal explosions, and the abandon ship order was given. Parker was among the 216 of nearly 3,000 sailors who did not make it off.

Born in Gary, Indiana, Parker graduated from Lincoln in May 1940 with degrees in math and sociology. He was "young, ambitious, buoyant with hope, exuberant with faith, looking to the future with serenity and courage," the *Clarion* said. He worked as a typist to pay for his school expenses and was part of Stagecrafters, the university band and the debate society.

Parker's first love was theater, and after graduating from Lincoln, he moved to Hollywood, California. While studying for a master's degree in sociology from the University of Southern California, he landed bit parts in several movies and socialized with many of the "current screen luminaries."

"He met a hero's death thousands of miles away from this quiet hill, in the flaming waters of the Coral Sea....He died young, but who can say his life was not complete? That it was not full and rich? He died for a dream, Democracy, as yet unrealized for all men....From that moment, the war at LU could no longer be discussed academically. Richard Parker's death 'in the line of duty' had brought the war's whole tragic import home to us," fraternity brother T. Thomas Fletcher wrote for the *Clarion*.

Staff Sergeant Maynard Harris, born in Festus, graduated from the Lincoln Lab High School in 1940. He was the second Lincolnite known to have been killed in action. In May 1943, Harrison was a warrant officer, serving in Tunisia with the 1963rd Ordnance Company (aviation), tactical and technical experts.

A popular poet and *Clarion* feature page editor, Jimmy Stewart was a tech sergeant from Evanston, Illinois. Stewart was killed in November 1945 in Calcutta, India, while driving an ambulance, which overturned and was burned during rioting. "Stewart became the first U.S. soldier to lose his life in the bloody Indian revolt against British domination," the *Clarion* reported. He was the last of four Lincolnites to be killed in action during World War II.

A part of the large spring call-up of Lincoln men to service, Stewart continued to write to his alma mater. While at Fort Belvoir, Virginia, he wrote, "What I learn now is not the French idioms nor the plant and animal species nor the prerequisites of a journalist. I am learning that there are actually grown men who cannot read nor write. I am learning how men react under strain."

Wendell Pruitt and Stewart had been charter members of the campus Stagecrafters. So, in celebration of the organization's tenth anniversary in April 1951, its performance of *Joan of Lorraine* was dedicated to them, with a special program published in their memory.

And for Veterans Day 1954, a portrait of Pruitt by J.S. Carpenter was presented to the alumni association by his mother, Melanie, on the steps of Young Hall.

In St. Louis, Pruitt was remembered as namesake for the first Black Sea Scout Ship in the St. Louis Council in May 1946 at St. Elizabeth Catholic Church, as well as AMVETS Post 41 and an elementary school. And in 1952, the largest housing development project at the time in St. Louis—a set of twenty eleven-story buildings with modern designs and fireproof brick—was named for him and Representative William Igoe. A 1940 Lincoln alumni, Eugene O. Bradley Jr., was appointed the first manager of the Wendell Oliver Pruitt Housing Unit.

Even the U.S. Air Force remembered Pruitt when it was looking for the name of a new training facility, although he was only a candidate.

Today, Pruitt's image is among the seventy-five African American aviators featured on a mural at the St. Louis–Lambert Airport. Spencer Taylor and Solomon Thurman completed the oil mural, *Black Americans in Flight*, in 1990. Pruitt's sister, Vesta, was president of the Committee for Aviation Mural Project Success and helped to raise $600,000 for the mural, including a $100,000 contribution from Anheuser-Busch. Everett Bratcher and Milton Lemmons are also featured in the mural.

In 1992, the same three Lincolnite Tuskegee Airmen were included in a St. Louis Science Center exhibit.

Chapter 13

DOUBLE VICTORY

Many of these Lincoln airmen broke racial barriers in military and civilian worlds. And others made an impact in their communities and the nation with their skills, knowledge and compassion. Their lives, by success and by character, helped gradually overcome racial mythology in their workplaces, communities and the nation.

CHARLES ANDERSON

Charles Anderson was one of the first Black Americans to earn a doctorate in meteorology from MIT in 1960, and he was the first tenured Black American professor at the University of Wisconsin in 1966. He later was professor of marine, earth and atmospheric sciences at North Carolina State University, where he was a major contributor to the severe storm forecasting program. Anderson, who is buried in Jefferson City, was an "internationally-known expert on the dynamics of severe storms, particularly tornadoes," according to the Raleigh, North Carolina *News and Observer* of October 26, 1994. His work was groundbreaking in the use of satellite data to detect cloud patterns and predict storms.

"I could have done engineering, but with my background in math and chemistry [I] seemed a good fit for meteorology," Anderson said in a 1992 interview with Earl Droessler.

Anderson developed an interest in tropical meteorology while at the Army Air Forces Meteorological Aviation Cadet Program at the University of Chicago. He put that on hold when he graduated in May 1943, as the army needed him at Tuskegee as a weather officer. After the war, the army sent Anderson to the Polytechnic University in Brooklyn to study high polymer chemistry.

At Douglas Aircraft Missiles and Space Systems Division, Anderson led the atmospheric analysis group. "I thought this was a great challenge, because it would put me into the space age and all the hoopla and glamour surrounding that new era," Anderson said. He also did pioneering work in the reduction of contrails of high-altitude jet aircraft.

During the 1970s at the University of Wisconsin, Anderson became chairman of African-American Studies. A coworker described Anderson as an "experienced veteran of the world [who] knew how to make it all survivable.... [Anderson] was knowledgeable, strategic and always saw where he was needed....He was savvy."

Charles E. Anderson became one of about forty Black weather officers with the U.S. Army Air Corps, instructing the Tuskegee Airmen pilots how to interpret weather conditions. After the war, he pioneered meteorological techniques and became an internationally known expert on severe weather prediction. *University of Wisconsin–Madison Archives.*

WALTER SANDERSON

Walter Sanderson moved from mechanical arts professor at Lincoln to superintendent of a metalworking business in Chicago in 1949. His friends said that "he took such an interest in meeting customers' needs for well-made, competitively-priced, delivered-on-time stampings, that his customers followed him when he started his own business," said the *Cardunal Free Press* of July 1, 1970.

The Walter H. Sanderson Manufacturing Company opened in 1965 on Chicago's South Side with a single press, and within six weeks, he needed to order another one. It manufactured automotive accessories, including head and tail lamps, directional signal housings, reflector bases and various tube parts, plus housings and steel castings for the electronics industry.

His company received Chicago's first "Project Own" loan for $240,000 from the Small Business Administration in August 1968. The government-private enterprise partnership was "aimed at speeding up the formation of minority-owned businesses," allowing him to expand operations and double employment.

The National Accelerator Laboratory in July 1970 awarded two contracts, valued at about $600,000, to Sanderson's company. These were the largest contracts awarded by the lab to a single minority entrepreneur up to that time in the United States. And their combined value was the largest contract awarded by the lab to a single supplier in the Greater Chicago area. "It is an unusual assignment, requiring great precision and there are only a few other companies in the country with experience in this field," the *Cardunal Free Press* reported.

The contracts were for 2 million laminations "to be used in the development of magnets for the largest single component—the main accelerator—of the 200 Billion Electron Volt proton synchrotron being built by the National Accelerator Laboratory" near Batavia, Illinois. When completed, the main accelerator was the "world's largest instrument of pure research," Sanderson said.

"His relatively small, but burgeoning company, is one of the fastest-growing stamping shops in the metropolitan area," the *Cardunal Free Press* said in 1970.

While president of Sanderson Industries, Sanderson served as president of the Chicago South Chamber of Commerce in 1974, when it received accreditation from the U.S. Chamber of Commerce. In 1985, he moved his operations to Atlanta, Georgia, being "less satisfied with Chicago business climate."

CLOVIS BORDEAUX

Clovis Bordeaux looked to education as a means for his future. Having earned dual bachelor's degrees at Lincoln for physics and math in 1939, he earned a master's of science in physics in June 1947 from St. Louis University, part of the archdiocese's integration efforts.

A great job with the U.S. Navy was waiting for him in Virginia. Borrowing a friend's car, Bordeaux took his new bride, Bernice, with him. Upon arrival, he asked about a place for his wife to stay. The navy office told him, "If they wanted him to have a wife, they would have issued him one," Jacques

Bordeaux recounted. So, the newlyweds turned down the job, turned around and headed west, this time to Chicago, Illinois.

Bordeaux encountered Lincoln fraternity brothers who helped him find housing and job opportunities. At the University of Chicago, he applied for the janitor's position, but his résumé ended up in the physics department. The academics were "dumbfounded the application belonged to a Black guy," his son, Jacques, said.

That fateful moment sent Bordeaux on a course to be "an African American scientist when there was not such a thing," the son said. A retired science instructor, Jacques Bordeaux said, "I appreciate Daddy's contribution; he would deny it. He was a humble guy."

And so, a Lincoln alumni moved into the pages of history, working on the cyclotron, one of the first particle accelerators, which opened the new study of nuclear science. Bordeaux worked with noted nuclear physicist Enrico Fermi under the University of Chicago's football field. "This was the place where the atom was split, and my dad was there," Jacques Bordeaux said.

The elder Bordeaux moved in 1961 to Southern California, which at this time "was a hotbed of physical science—rockets, propulsion, guidance." Defense contracts were "flowing like water," and Bordeaux took a job designing ground system project components with Howard Hughes Aircraft. In the 1960s, when there were few Black engineers, Bordeaux was a supervising physicist for an internationally renowned aerospace firm. "The guys from Hughes Aircraft said he was the most authentic man they ever met, though he had seen bigotry and experienced deep pain," Jacques Bordeaux said.

This led Bordeaux to another unique opportunity. Pope Paul VI made an unprecedented visit to Bogotá, Colombia, in 1968 to seek social reform. As the section head of the Hughes ground systems division and lead engineer on that transmitter, Bordeaux and his team flew to Bogotá, towed the transmitter to the highest point in the Andes Mountains, tested it and had it ready to transmit the pope's visit.

Remembering his alma mater and wanting to encourage young Black scientists, Bordeaux coordinated a matching scholarship with Hughes Aircraft Company for a physics student at Lincoln. The first recipient was named in April 1980. After retirement, Bordeaux continued to teach math and science basics to Black college students.

From substandard education, Bordeaux achieved his own type of excellence, navigating social, educational, professional and military challenges. He was remembered for being a good communicator, an authentic person and having precision in his speech and manner.

Richard Pullam

Unlike most of the Tuskegee Airmen from Lincoln, Richard Pullam remained on active duty after the war. When the U.S. Air Force was organized in 1947, the 332nd Fighter Group was reorganized and the 477th Composite Group was deactivated. All Black aviation personnel were moved to Lockbourne (Ohio) Army Air Force base, where they were equipped with P-47 Thunderbolts.

Pullam, who had been an instructor at Tuskegee at the end of the war, rose to commander of the 301st Fighter Squadron in 1948, although the promotion was short-lived.

The 332nd Fighter Group was "outstandingly good" in its performance at a 1948 war games simulation in Florida, the largest since the war, according to Major General William Old, commander of the 9th Air Force. He commended the 332nd, commanded by Colonel Benjamin Davis Jr., on its dive bombing, rocket firing and strafing. The 332nd also won the U.S. Air Force's first aerial gunnery competition in Las Vegas, Nevada. "It was the 332nd Fighter Group's last hurrah," historian Daniel Haulman said.

In July 1948, President Harry Truman signed Executive Order 9981, mandating the U.S. Armed Services desegregate. The 332nd group and its squadrons were deactivated and personnel reassigned to formerly all-white units as the branch integrated.

In June 1949, Pullam was assigned to Keesler Army Air Field, which became the branch's electronics training center, and the 3380th Technical Training Group operated the school. Within six months, Pullam was released from service, although he remained with the air force reserve until 1971. "To hear my grandfather talk about his time as a Tuskegee Airman was amazing. He would tell you what books couldn't tell you," said Sharice Livingston, granddaughter of Pullam's ex-wife.

Pullam became a college professor and a founding member of the Heart of American Chapter of the Tuskegee Airmen Inc., which created a $1,500 scholarship for thirty-five students interested in aeronautics each year.

Although he retired to his family cabin on Lake Placid Resort, near Stover, Pullam still followed aviation. "I wish I could fly those new planes. I love speed," Pullam told the *Sedalia Democrat*.

Lake Placid was a fifteen-acre resort created in 1937 as "a recreational center for colored people in the Missouri Ozarks," historian Gary Kremer said in *Race and Meaning*. One of the first six cabins for Black professionals from Kansas City was called "My Blue Heaven." Owned by pharmacist

William Houston, who died in 1941, it passed to Pullam's father, Arthur "Chic" Pullam Sr., Houston's half-brother, who was a postal worker and professional baseball player.

"It was quite prestigious to have a cabin at Lake Placid during the 1940s," Chic's daughter and Pullam's sister, Barbara, recalled. "My Blue Heaven" passed to the Tuskegee Airman in 1968, at his father's death. There, he lived year-round.

Pullam died of a heart attack in 1997, the last combat pilot member of the Heart of America Chapter of the Tuskegee Airmen Inc., and was buried at Leavenworth National Cemetery. "He was a hero, but you would never have known he was a hero. He didn't act like he was a hero. He was very down-to-earth. He laughed, he loved life. He was brave, but modest," said the Tuskegee Airmen chapter president Vivian Bassa.

ALBERT GAINES

Like Pullam, Albert Gaines remained in the service. He progressed to the rank of captain and eventually became a flight instructor. "He successfully integrated the New York National Guard," the Syracuse University Office of Veterans and Military Affairs said. And "with support from important figures, Gaines was accepted into and attended Syracuse University—it was a 'Jackie Robinson moment,' as he called it."

Gaines joined IBM in 1956 in Kingston, New York, after setting the highest opening exam score on record. He advanced from engineering systems trainee to supervisor of an electronics group working on rockets, then from diagnostic engineer to diagnostic programmer. He worked more than thirty-five years for IBM, as an attorney and computer engineer.

In November 1969, Gaines was "on loan" to the Urban League of Westchester as director of the Small Business Advisory Council, whose purpose was to help minority group members find and successfully operate their own businesses. The project was based in Peekskill, where Gaines lived, and was involved with police personnel on juvenile delinquency problems and the NAACP dealing with education and mental health. In 1974, Gaines was named Man of the Year by the Peekskill Chapter of the NAACP.

Gaines "served as an advocate for civil rights and held significant positions in several local political committees, state associations and in Westchester community organizations; but most importantly, Gaines was a diverse, kind-hearted and intelligent man….His lifetime achievement was helping race relations in this nation. However…this humble and discrete man never spoke of his accomplishments," according to Syracuse University's alumni webpage.

Stewart Fulbright Jr.

Stewart Fulbright Jr. earned the rank of captain before his discharge and remained in the air force reserve through 1968.

A French major, teaching the language at Lincoln for two semesters, Fulbright used his language skills little after the war. Instead, he leaned into his minor at Lincoln in business administration. He earned a Master of Business Administration from the University of Chicago. In 1947, he began teaching commerce at North Carolina Central University, Durham, while earning his doctorate degree from The Ohio State University. He was awarded postdoctoral work in 1962 through the Danforth fellowship at the University of Pennsylvania.

Fulbright remained active in the Lincoln University alumni association. "I love Lincoln and I realize just how important Lincoln was in my life, and the preparation that it gave me," he said.

Fulbright advanced to dean of the undergraduate school and campus representative to the Woodrow Wilson Foundation, while helping incorporate the North Carolina Manpower Development Corporation.

In August 1968, he was named the first chairman of the business and economics department, which he had helped to create. At the same time, he became the second Black man named to the U.S. Department of Agriculture's Commodity Credit Corporation's five-man advisory board, which met four times a year to survey the corporation policies, including price support and production stabilization.

Fulbright was among the three hundred Tuskegee Airmen who received the Congressional Gold Medal from President George W. Bush in March 2007. With him was his daughter, Gina Fulbright-Powell, who said that her father was pleased with the recognition sixty years after his service. At the same time, it was disappointing that the Tuskegee Airmen initially were asked to pay for their own medals, although that was corrected later. "Without a doubt, white airmen never would have [been asked]."

"My father was a very modest man and believed in doing your best and letting the results speak for themselves," his son, Ed, told the *Springfield*

(MO) News Leader when his father was posthumously inducted into the Springfield Public Schools Hall of Fame.

JOHN HUGHES

John Hughes, member of the first Lincoln CPT class, taught physical education for thirty-seven years in the Springfield (Missouri) School District. He was in the Prince Hall Masons, elected as grand master of the Grand Lodge of Missouri in 1965 and became a 33rd degree Mason.

Hughes also was active in community civil rights and politics, serving on the Springfield Area Law Enforcement Council. He was vocal about "a double standard operating in Springfield and in the nation." As part of a 1969 speakers panel for the Drury College Crime and Corrections Institute, Hughes said, "There has always been a double standard. I know it's true and you know it's true, and there will be 'til men and women come together under one God and treat people as they wish to be treated."

As president of the Springfield NAACP chapter in 1976, he pointed out a communications gap, not between city hall and the Black community but rather within the city government. He also led the Prince Hall Grand Lodge sponsorship of low-income housing, including a 138-unit apartment group in Springfield named for him.

Hughes "served as a role model for young blacks in Springfield, encouraging them to complete their educations and develop their talents, and was a tireless advocate for low-income people and the elderly.... [At his death, he was] described by those who knew him as a 'wonderful person' who was consistently upbeat, charitable and involved in the community," the *Howard-News* reported.

JULIAN WITHERSPOON

As a social welfare caseworker in Wayne County, Michigan, in August 1965, Julian Witherspoon accused Mayor Jerome Cavanaugh of "bureaucratic control of the city's anti-poverty program." The *Los Angeles Times* said in 1965, "A year ago, his voice would have carried little farther than the doorway of his $70-a-month apartment in a four-family flat in a poverty-ridden, predominantly Negro neighborhood. Today, elected by his neighbors to sit on the highest anti-poverty advisory council in the city, Witherspoon stridently speaks his piece....He takes home less than $4,000/year to care for a family of six, however, his income was considered too high for him to be deemed 'poor,' though his choice to live among the poor qualified him to speak for them."

The *Detroit Free Press* described Witherspoon as "a classic man-in-the-middle. To city hall he's a gadfly, with an annoying indifference to bureaucratic protocol and political realities. To many people on the street, he is a man with good ideas but no money to back them up."

In the summer of 1967, when the nation's largest civil disturbance erupted there, Witherspoon was an "organizer of block clubs throughout Detroit," chairman of the Detroit Inner City Voters League and an anti-poverty program official in the 12th Street area.

"Nobody wants to listen to the gang leaders, the malcontents. Yet, those are the ones who can call a meeting and get the cats out there," Witherspoon told the *St. Louis Post-Dispatch*. "This bitterness has got to be assuaged somewhere, or worse conflagrations will come. Telling it like it is means acknowledging this bitterness and really responding to it at every level of our society and from every ethnic group....We've heard our parents say 'it will be better in your time,' but here I am in the same box....I have got to stand with this movement, otherwise I am standing for a society that is wrong."

WILLIAM CRUMP

From Lincoln's final CPT class, William Crump continued his education at Northwestern University, studying the value of interracial workplaces. He interviewed employees and employers from nineteen interracial workplaces and fifteen that were Black-only. "Both employees and employers believe Negroes are better producers in inter-racial rather than in segregated work units....Competition between the two races serves to stimulate personal endeavors," the survey revealed.

Crump joined the faculty at Tennessee Agricultural and Industrial State College, Nashville, Tennessee, in 1950, after earning a doctorate in business administration. He moved from director of the business school within the education department to head of the university public relations department.

In 1982, Crump was listed in the *Who's Who in Black Corporate America*, when he was dean of the college of business and public administration at the University of the District of Columbia, Washington, D.C. In 1991, he published the 713-page *The Story of Kappa Alpha Psi*.

ERNEST BOONE III

Ernest Boone III was a high school teacher before serving as an administrative specialist in the U.S. Army Air Corps. Afterward, he played football for the University of Michigan and earned his master's degree from Butler University. After his service, he was one of Columbia's first two Black police officers, appointed in April 1946. He moved to Indianapolis, Indiana, where he was an instructor in Army Finance at Fort Harrison and then became a teacher and later principal in the public schools.

His father, Ernest Boone Jr., was active in the state Democratic Party, traveling the state on behalf of candidates to lobby between the African American community and the political leaders. The CPT cadet's grandfather was principal at the Dunbar School and later appointed to the Tipton School for Girls.

BERTRAN WALLACE

Bertran Wallace also made the military a career, being laid to rest at Arlington National Cemetery. He returned from Europe, serving as an air corps military policeman, decorated with two Bronze Stars for the Naples-Foggio and Rome-Arno campaigns.

The transition to integration of the military was difficult on the first Black officers, like Wallace. "Guys like us paved the way. We...had to go through the 'boys,' the 'niggers,' the 'coons' and the 'eenie-meanie-miney-mo,' and we just proved that we were equal to them or better than they were."

"Serving in two wars, plus almost 30 years of peacetime duty, didn't take the edge off Bertran Wallace, a proud man who fought as hard against prejudice within the Army as he did against the Germans in North Africa and Italy and the North Koreans. Wallace's fondest memories are from World War II, when his all-black MP unit served with distinction after the Allies captured Rome." His story is preserved in the Library of Congress' Veterans History Project.

He worked as a printer for several newspapers, including the *New York Times* and the *New York Herald Tribune*, while he earned his master's degree in 1948 from the teachers' college at Columbia University, New York.

In 1949, Wallace was commissioned a second lieutenant in a post-integration army. He was assigned as a staff specialist, based on his

Bertran Wallace served more than thirty years, fighting two wars and a constant battle against prejudice. He is buried at Arlington National Cemetery. *Library of Congress.*

graphic arts career, with others who were in the news and media. At his first meeting with his new unit, however, he was met by a lieutenant colonel who saw a Black man with gold bars on his uniform and asked, "What the hell are you doing here?!" Wallace said, "I responded as I normally do when questions are asked like that. I came back, I guess, rather sharply and said, 'Because I belong here.'"

About two weeks later, he received a "twix" from the governor saying "either you transfer to military police or resign your commission." Wallace said he did not understand why. "I was a staff specialist. That's what I was commissioned on. And I had my degree from Lincoln, and I had my graduate work from Columbia, and I was working for the newspaper. So, I was very qualified." A lawyer later told him that he could have sued the federal government for that. "But, because I believe in my country and because I believe that whatever I accomplish, that I must merit that, I wasn't going to use the law to fight my problems," Wallace said.

Wallace accepted the transfer, although he did not want to be a military police officer again. The 33rd Military Police Battalion had one Black company and three companies that were white only. "Company D was the best company in the battalion. We knew that we were segregated, so we wanted to prove to others that we were the best."

As the Korean War developed, his unit was repeatedly put on and off deployment alert. "I got tired of [it]. So, I volunteered for the Korean War," Wallace said.

Despite his commission as an officer and his experience, Wallace again was relegated to the military police. Years later, he learned that a white superior officer "had written in my report that I was unfit to be an officer and that I needed to enroll in a course of Americanization."

"I had grown up in the South and I just couldn't handle prejudice very well. I discovered that when I was overseas in North Africa and Italy....I think that all human beings are equal in the eyesight of God...that's my strength, and that's my belief."

Wallace was assigned to military police school in Georgia. "There was a lot of problems there that I encountered with officers, prejudice." Yet he finished ninth out of eighty-eight officers. Then he was sent overseas—the only officer of color in his unit.

Although President Harry Truman signed the integration order, "still people had their hidden agendas and didn't like me because of the color of my skin." Wallace's ancestry included a maternal grandfather who was Mexican; his paternal grandmother was Indian, and his paternal grandfather was Scottish. "So, what does that make me? Anyhow, I have this complexion and I'm very proud of it because that's the only one that I know."

He was assigned to Koju-do, a POW camp with a troubled reputation. Of more than four thousand officers from various branches, only six were not white. Wallace served as commander over five compounds, thirteen thousand prisoners and thirty-two war criminals. Eventually, he was made an intelligence officer and became deputy provost marshal, "a big step," he said. When his eleven-month tour ended, he was the oldest officer there. One of his responsibilities was to meet daily with the senior colonel of the POWs to "negotiate." Because he was an academic and an athlete, the Koreans respected him.

"I was considered a very good soldier because I cared about the men and the men knew that," Wallace said. Although he didn't receive medals or ribbons, like the younger white officers who followed him, Wallace said that he was pleased to be invited to socialize with the enlisted men. "Anytime enlisted men invite an officer to socialize with them, you know you must bring something to the table."

He was often a pioneer, such as being the first Black instructor at the military police school in Little Rock, Arkansas. But that often came with inescapable racism. Ultimately, he left the military because of the "overt prejudice."

Returning to the print industry and civilian life, Wallace became a teacher at the New York School of Printing in 1960. "The idealism in me and my faith in man/womankind is what led me to teach. I wanted to do something with my life to make a contribution. I felt God had granted me many talents, and I needed to share them."

Wallace went on to become director of several vocational education departments in New York state school districts and a professor at New Jersey State College.

As president of the Yonkers NAACP chapter in 1987, he led the chapter's seven-year desegregation suit against the city, which intentionally segregated minorities for four decades. When the 2nd U.S. Circuit Court of Appeals upheld the landmark racial discrimination ruling in December 1987, Wallace said "Hot dog! Beautiful. Terrific. Bravo. It means justice will be done. We will continue with the fight and the struggle."

The most important thing to Wallace was "people. It's my mission to bring about some kind of hope and peace and affection for mankind. My goal? To make people smile and love and be friendly."

SOURCES

Chapter 1

Archives yearbook. Lincoln University (Missouri) Archives, 1940, 1941.

Aviation Heritage Park. "Willa Brown." www.aviationheritagepark.com/ aviators/willa-brown.

The Cauldron yearbook. Cambridge: Massachusetts Institute of Technology, 1932.

Francis, Charles. *The Tuskegee Airmen: The Men Who Changed a Nation.* N.p.: Branden Publishing Company, 1997.

Holland, Antonio F. *Soldiers' Dream Continued: A Pictorial History of Lincoln University of Missouri.* N.p.: Lincoln University Printing Services, 1991.

Homan, Lynn, and Thomas Reilly. *Black Knights: The Story of the Tuskegee Airmen.* Charleston, SC: Arcadia Publishing, 2001.

Lincoln University (Missouri) Archives. Vertical file.

Missouri State Penitentiary Register. Missouri State Archives.

Naomi Long Madgett Collection. News clipping.

National Air and Space Museum. "Wall of Honor: Charles Malcolm Ashe." https://airandspace.si.edu/support/wall-of-honor/charles-malcolm-ashe.

National Museum of the United States Air Force. "Civilian Pilot Training Program." www.nationalmuseum.af.mil/Visit/Museum-Exhibits/Fact-Sheets/Display/Article/196137/civilian-pilot-training-program.

National Park Service. "Moton Field." https://www.nps.gov/museum/ exhibits/tuskegee_airmen/moton_field.html.

Scott, Lawrence, and William Womack. *Double V: The Civil Rights Struggle of the Tuskegee Airmen.* East Lansing: Michigan State University Press, 1998.

U.S. Census, 1930, 1940.

U.S. Congress, Senate hearings. Committee on Naval Affairs, 1939.

Atlanta (GA) Constitution. June 18, 1939.

Chicago (IL) Tribune. June 13, 1942.

Ebony. March 1972.

Indianapolis (IN) News. October 18, 1956.

Indianapolis (IN) Star. May 5, 1945.

Jet. September 10, 1964.

Lincoln University (MO) Clarion. March 19, 1943; November 20, 1942; November 9, 1940; December 20, 1940; February 13, 1953; October 23, 1942.

Montgomery (AL) Advertiser. May 24, 1935.

New York (NY) Age. October 12, 1940; February 19, 1938; September 28, 1940; November 16, 1940.

Pittsburgh (PA) Courier. November 4, 1939; September 27, 1941; May 2, 1942.

St. Louis (MO) Post-Dispatch. December 15, 1940; May 27, 1938.

Chapter 2

Braafladt, Kevin. "The Story of the Only Regiment Commanded Entirely by Black Officers during World War I." United States Army, January 28, 2021. https://www.army.mil.

National Court Reporters Association. Bertran F. Wallace Collection, 1942. Personal Narrative. Library of Congress. https://www.loc.gov/item/afc2001001.01649.

Alabama Tribune (Montgomery). May 25, 1951.

Atchison (KS) Daily Globe. November 11, 1954.

Daily Press (Newport News, VA). October 25, 1970.

Lincoln University (MO) Clarion. October 13, 1944; October 12, 1945; February 14, 1947; November 19, 1954.

New York (NY) Age. May 17, 1941; March 29, 1941.

Pittsburgh (PA) Courier. June 10, 1944.

Sedalia (MO) Weekly Democrat. May 26, 1939.

Selma (AL) Times-Journal. May 24, 1953.

Spokane (WA) Chronicle. May 21, 1969.
Times Herald (Port Huron, MI). December 30, 1969.

Chapter 3

Archives yearbook. Lincoln University (Missouri) Archives, 1942.

Barbour, George Edward. "Early Black Flyers of Western Pennsylvania, 1906–1945." Penn State Libraries Open Publishing. journals.psu.edu.

Committee on Naval Affairs. Senate hearings, U.S. Congress, 1939.

Fold3. United States veterans' grave sites, circa 1775–2019. Fold3.com.

Fontenot, Renee, Major. "Upward Mobility: The Civilian Pilot Training Program, War, and Society in the American Century." School of Advanced Air and Space Studies, Air University, Maxwell Air Force Base, Alabama, 2018.

Francis, Charles. *The Tuskegee Airmen: The Men Who Changed a Nation.* N.p.: Branden Publishing Company, 1997.

Haulman, Daniel. *The Tuskegee Airmen Chronology.* Athens, GA: NewSouth Books, 2018.

Kraus, Theresa L. "The CAA Helps America Prepare for World War II." Federal Aviation Administration. https://www.faa.gov.

National Museum of the U.S. Air Force. nationalmuseum.af.mil.

National Park Service. "Moton Field." https://www.nps.gov/museum/ exhibits/tuskegee_airmen/moton_field.html.

We Gotta Story to Tell. "Tuskegee Airmen: Richard 'No Middle Name' Davis." Podcast. African American Museum of Southern Arizona, May 24, 2022.

Birmingham (AL) News. September 3, 1939.
The Black Dispatch (Oklahoma City, OK). September 5, 1942.
The Crisis (New York). March 1943.
Jefferson City (MO) Post-Tribune. March 8, 1938; August 3, 1929; December 22, 1929; November 25, 1940; August 5, 1941; September 19, 1941; November 16, 1994; May 20, 1977.
Lincoln University (MO) Clarion. April 30, 1943.
New York Age. August 23, 1941.
Pittsburgh (PA) Courier. August 9, 1941.
Pittsburgh (PA) Press. July 12, 1942.
St. Louis (MO) Post-Dispatch. May 4, 1992.

Chapter 4

Appel, Susan. "Jefferson City Community Center." National Register nomination form, National Park Service, 1992.

Archives yearbook, Lincoln University (Mo.) Archives, 1942.

"Black Paris Tours" Facebook page, April 18, 2023.

Fold3. United States veterans' grave sites, circa 1775–2019. Fold3.com.

Francis, Charles. *The Tuskegee Airmen: The Men Who Changed a Nation.* N.p.: Branden Publishing Company, 1997.

Haulman, Daniel. *The Tuskegee Airmen Chronology.* Athens, GA: NewSouth Books, 2018.

Holland, Antonio F. *Soldiers' Dream Continued: A Pictorial History of Lincoln University of Missouri.* N.p.: Lincoln University Printing Services, 1991.

Pasadena (CA) city directory, 1969.

Pisano, Dominick. "Eugene J. Ballard." Smithsonian National Air and Space Museum, October 12, 2010.

Schwerin, Mark. "Tuskegee Airmen Get Delayed Salute." *Battle Creek (MI) Enquirer*, March 1, 1998.

Smith, Walker. *Mello Yello: The Incredible Life Story of Jack the Rapper.* N.p.: Sonata Books, 2015.

U.S. Census, 1930, 1940.

U.S. Department of Veterans Affairs, BIRLS Death File, 1850–2010. Via Ancestry.com.

United States World War II Army Enlistment Records, 1938–1946. Via Ancestry.com.

We Gotta Story to Tell. "Tuskegee Airmen: Richard 'No Middle Name' Davis." Podcast. African American Museum of Southern Arizona, May 24, 2022.

Cardunal Free Press (Carpentersville, IL). July 1, 1970.

The Crisis (New York). March 1943.

Daily Capital News (Jefferson City, MO). November 20, 1942; December 25, 1942.

Leavenworth (KS) Times. August 15, 1977.

Lincoln University (MO) Clarion. August 8, 1941; October 23, 1942; October 30, 1942; November 6, 1942; November 13, 1942; November 20, 1942; November 27, 1942; December 4, 1942; February 5, 1943; May 7, 1943; May 21, 1943; May 28, 1943; December 15, 1944.

New York Age. July 3, 1943.

Pittsburgh (PA) Courier. December 3, 1943.

Chapter 5

Appel, Susan. "Jefferson City Community Center." National Register nomination form, National Park Service, 1992.

"Black Paris Tours" Facebook page, April 18, 2023.

Brooks, Michelle. *Interesting Women of the Capital City*. N.p.: Kindle Direct Publishing, 2021.

Fold3. United States veterans' grave sites, circa 1775–2019. Fold3.com.

Francis, Charles. *The Tuskegee Airmen: The Men Who Changed a Nation*. N.p.: Branden Publishing Company, 1997.

Haulman, Daniel. *The Tuskegee Airmen Chronology*. Athens, GA: NewSouth Books, 2018.

Pasadena (CA) city directory, 1969.

Pisano, Dominick. "Eugene J. Ballard." Smithsonian National Air and Space Museum, October 12, 2010.

Schwerin, Mark. "Tuskegee Airmen Get Delayed Salute." *Battle Creek (MI) Enquirer*, March 1, 1998.

U.S. Census, 1930, 1940.

Cardunal Free Press (Carpentersville, IL). July 1, 1970.

Daily Capital News (Jefferson City, MO). November 20, 1942; December 25, 1942.

Lincoln University (MO) Clarion. August 8, 1941; October 23, 1942; October 30, 1942; November 6, 1942; November 13, 1942; November 20, 1942; November 27, 1942; December 4, 1942; February 5, 1943; May 7, 1943; May 21, 1943; May 28, 1943; December 15, 1944.

New York Age. July 3, 1943.

Chapter 6

Archives yearbook. Lincoln University (Missouri) Archives, 1940.

The Bulletin. Lincoln University of Missouri, 1942–43.

Fold3. United States veterans' grave sites, circa 1775–2019. Fold3.com.

Francis, Charles. *The Tuskegee Airmen: The Men Who Changed a Nation*. N.p.: Branden Publishing Company, 1997.

Hoard, Adrienne. Interview with author, November 22, 2022.

Jackson, Richard A. "Sport Nianianis." Fall 1941. Adrienne Hoard Collection.

Naval History and Heritage Command. "African Americans and the Navy: WWII." https://www.history.navy.mil/our-collections/photography/people---special-topics/african-americans-in-the-navy/african-americans-and-the-navy--wwii.html.

Stewart B. Fulbright Collection. Veterans History Project, American Folklife Center, Library of Congress, 2001.

U.S. Census, 1930.

U.S. Department of Veterans Affairs, BIRLS Death File, 1850–2010. Via Ancestry.com.

Who's Who in Colored America. New York, 1950.

Jefferson City (MO) News Tribune. December 10, 1989.

Lincoln University (MO) Clarion. December 11, 1942; February 5, 1943; February 19, 1943; March 5, 1943; March 12, 1943; March 19, 1943; March 26, 1943; April 23, 1943; April 30, 1943; October 8, 1943; November 12, 1943; May 7, 1975; May 14, 1943; May 21, 1943; May 28, 1943; April 27, 1956.

Moberly (MO) Monitor-Index. January 21, 1942; September 8, 1943.

New York Age. March 6, 1943.

Springfield (MO) News-Leader. October 14, 2012.

St. Clair (MO) Chronicle. May 6, 1943.

St. Louis (MO) Star Times. April 16, 1945.

Sunday News and Tribune (Jefferson City, MO). February 16, 1947.

Chapter 7

Amick, Jeremy Paul. *Together as One: The Legacy of James Shipley, World War II Tuskegee Airman.* N.p.: Yorkshire Publishing, 2018.

AMVETS Post 41. https://www.angelfire.com/mo/AMVETSPOST41/index.html.

Archives yearbook. Lincoln University (Missouri) Archives, 1936, 1940, 1941.

Bordeaux, Jacques. Interview with author, 2019.

Bratcher Scrapbook Collection. Missouri History Museum Archive.

Cecil Peterson Collection on the Tuskegee Airmen. Special Collections & University Archives, University of California–Riverside.

Dryden, Charles W. *A-Train: Memoirs of a Tuskegee Airman.* Tuscaloosa: University of Alabama Press, 2002.

Find-A-Grave.

Francis, Charles. *The Tuskegee Airmen: The Men Who Changed a Nation.* N.p.: Branden Publishing Company, 1997.

Hanchett, Tom. "1713 Madison Avenue." McCrorey Heights. History South. historysouth.org.

Haulman, Daniel. *Eleven Myths About the Tuskegee Airmen.* Athens, GA: NewSouth Books, 2012.

———. *The Tuskegee Airmen Chronology.* Athens, GA: NewSouth Books, 2018.

Homan, Lynn, and Thomas Reilly. *Black Knights: The Story of the Tuskegee Airmen.* Charleston, SC: Arcadia Publishing, 2001.

Krewson, Brittany. "St. Louis' Red-Tail Angels." Missouri Historical Society, July 30, 2018. mohistory.org/blog.

National Archives and Records Administration.

National Park Service. "Moton Field." https://www.nps.gov/museum/exhibits/tuskegee_airmen/moton_field.html.

Scott, Lawrence, and William Womack. *Double V: The Civil Rights Struggle of the Tuskegee Airmen.* East Lansing: Michigan State University Press, 1998.

Stewart B. Fulbright Collection. Veterans History Project, American Folklife Center, Library of Congress, 2001.

Theroff, Anna. "Judge Ernest Moss Tipton." Missouri Courts, 2019. www.courts.mo.gov.

Tuskegee Army Air Field. Report of Aircraft Accident, May 15, 1944. National Archives and Records Administration.

United States World War II Army Enlistment Records, 1938–1946. Via Ancestry.com.

U.S. Department of Veterans Affairs, BIRLS Death File, 1850–2010. Via Ancestry.com.

Woloszyn, Marisa. "The Story of Charles E. Anderson and the Tuskegee Airmen." *TMJ4*, Milwaukee, Wisconsin, February 21, 2021.

Associated Press. January 17, 1941.

Daily Item (Sunbury, PA). March 24, 2011.

Daily Sentinel (Grand Junction, CO). January 28, 1945; January 30, 1946; March 5, 1946; December 30, 1986.

Lincoln University (MO) Clarion. October 30, 1942; November 6, 1942; November 20, 1942; February 5, 1943; February 12, 1943; February 19, 1943; March 5, 1943; April 9, 1943; May 7, 1943; May 21, 1943; November 5, 1943; November 19, 1943; November 26, 1943; December 10, 1943; December 15, 1944.

Moberly (MO) Monitor-Index. September 13, 1943; January 23, 1967.

New York Age. November 14, 1942.
Pittsburgh (PA) Courier. January 2, 1943; February 13, 1943; March 6, 1943; August 21, 1943; September 11, 1943.
Rantoul (IL) News-Gazette. October 31, 1941.
St. Louis (MO) Post-Dispatch. December 4, 1960; March 24, 2011.

Chapter 8

Bordeaux, Jacques. Interview with author, 2019.
Dryden, Charles W. *A-Train: Memoirs of a Tuskegee Airman.* Tuscaloosa: University of Alabama Press, 2002.
Fold3. Missing air crew report. Fold3.com.
Francis, Charles. *The Tuskegee Airmen: The Men Who Changed a Nation.* N.p.: Branden Publishing Company, 1997.
Ganett News Service. Clipping, February 13, 1998. Naomi Long Madgett Collection.
Haulman, Daniel. *The Tuskegee Airmen Chronology.* Athens, GA: NewSouth Books, 2018.
Jefferson, Alexander. *Red Tail Captured, Red Tail Free: Memoirs of a Tuskegee Airman and POW.* New York: Fordham University Press, 2017.
Madgett, Naomi Long. Interview with author, November 4, 2019.
———. *Pilgrim Journey.* Detroit, MI: Wayne State University Press, 2006.
National Park Service. "Moton Field." https://www.nps.gov/museum/exhibits/tuskegee_airmen/moton_field.html.
100th Bomb Group Foundation. 100thbg.com.
Tuskegee Airmen, Hugh J. White Chapter. tuskegeeairmenstl.com.
World War II Reports of Separation. Missouri State Archives.

Jefferson City (MO) Lincoln Clarion. October 15, 1943.
Pittsburgh (PA) Courier. October 2, 1943.
St. Louis (MO) Post-Dispatch. September 26, 2009.

Chapter 9

Amick, Jeremy Paul. *Together as One: The Legacy of James Shipley, World War II Tuskegee Airman.* N.p.: Yorkshire Publishing, 2018.

AMVETS Post 41. https://www.angelfire.com/mo/AMVETSPOST41/index.html.

Campbell, Crispin. "Black Pilots of '40s Charted New Horizons." *Washington Post*, February 23, 1983.

Cecil Peterson Collection on the Tuskegee Airmen. Special Collections & University Archives, University of California–Riverside.

Commemorative Air Force. "Profiles of Tuskegee Airmen: Lee Archer." commemorativeairforce.org.

Francis, Charles. *The Tuskegee Airmen: The Men Who Changed a Nation*. N.p.: Branden Publishing Company, 1997.

Hartzer, Ronald B. "A Look Back at Black Aviation Engineer Units of World War II." *Webster (MO) Advertiser*, November 9, 1967.

Haulman, Daniel. *The Tuskegee Airmen Chronology*. Athens, GA: NewSouth Books, 2018.

Huntley, Craig. E-mail correspondence with author.

Malone, Ross. *Missouri's Forgotten Heroes*. N.p.: CreateSpace Independent Publishing, 2016.

National Archives and Records Administration. Airmen personnel records.

National Park Service. "Moton Field." https://www.nps.gov/museum/exhibits/tuskegee_airmen/moton_field.html.

New Orleans Volunteer Association. "Obituaries Orleans Parish Louisiana." September 2006. files.usgwarchives.net.

Peirson, Scot. "Destroyer Sunk by Tuskegee Airmen—More Information Needed." Post, July 23, 2007. ww2aircraft.net.

Scott, Lawrence, and William Womack. *Double V: The Civil Rights Struggle of the Tuskegee Airmen*. East Lansing: Michigan State University Press, 1998.

Sekhon, Sharon, et al. "Shouting from the Margins: Black Orange County, 1960 to 1979, Project." Juel Farquhar (1922–) section.

Stanley, Sandler. *Segregated Skies*. Washington, D.C.: Smithsonian, 1992.

World War II Reports of Separation. Missouri State Archives.

The Gazette (Cedar Rapids, IA). September 6, 1942.

Lincoln University (MO) Clarion. October 8, 1943; October 6, 1944; December 15, 1944; January 12, 1945; January 19, 1945; April 13, 1945; April 29, 1945; April 27, 1945.

New York Age. February 3, 1945.

Pittsburgh (PA) Courier. November 4, 1944.

Sedalia (MO) Democrat. March 22, 1996.

St. Louis (MO) Post-Dispatch. November 28, 1944; November 29, 1944; December 6, 1944; December 12, 1944; April 19, 1945.
St. Louis (MO) Star-Times. November 27, 1944; April 16, 1945.

Chapter 10

Dama, Valentine. KJLU. Interview with Dr. Stewart Fulbright, May 7, 2010.
477th Fighter Group. "History." www.477fg.afrc.af.mil.
Haulman, Daniel. *The Tuskegee Airmen Chronology.* Athens, GA: NewSouth Books, 2018.
Lincoln University. "Alumni Line." 2012.
Moye, J. Todd. *Freedom Flyers: The Tuskegee Airmen of World War II.* New York: Oxford University Press, 2012.
Stewart B. Fulbright Collection. Veterans History Project, American Folklife Center, Library of Congress, 2001.
U.S. Department of Veterans Affairs, BIRLS Death File, 1850–2010. Via Ancestry.com.
U.S. Select Military Registers, 1951. Via Ancestry.com.
White, Gerald A., Jr. "Tuskegee (Weather) Airmen: Black Meteorologists in World War II." Air Power History, Summer 2006.

Buffalo Evening News. February 3, 1945.
Lincoln University (Jefferson City, MO) Clarion. December 20, 1940.
New York Age. September 2, 1944.
Pittsburgh (PA) Courier. December 18, 1943; July 22, 1944.
Springfield (MO) News-Leader. October 14, 2012.

Chapter 11

U.S. Department of Veterans Affairs, BIRLS Death File, 1850–2010. Via Ancestry.com.
Find-A-Grave.
Fold3. United States veterans' grave sites, circa 1775–2019. Fold3.com.
World War II Reports of Separation. Missouri State Archives.

The Crisis. June/July 1952.
Lincoln University (Jefferson City, MO) Clarion. March 19, 1943; May 28, 1943; October 27, 1944; March 30, 1945.
Pittsburgh (PA) Courier. May 17, 1952.

Chapter 12

Find-A-Grave.

Fold3. United States veterans' grave sites, circa 1775–2019. Fold3.com.

History. "An Ammunition Ship Explodes in the Port Chicago Disaster." This Day in History. www.history.com.

Hoard, Adrienne. Interviews with author.

Lincoln University (Missouri) Archives. Vertical file.

National Archives and Records Administration. Personnel records.

U.S. Census, 1940.

U.S. Department of Veterans Affairs, BIRLS Death File, 1850–2010. Via Ancestry.com.

World War II Reports of Separation. Missouri State Archives.

Daily Capital News (Jefferson City, MO). April 22, 1964; July 7, 1977.

Jefferson City (MO) Post-Tribune. February 12, 1961; May 18, 1962; January 17, 1969; January 20, 1969; May 9, 1973.

Lincoln University (Jefferson City, MO) Clarion. November 27, 1942; March 5, 1943; March 12, 1943; March 19, 1943; April 3, 1943; May 14, 1943; May 28, 1943; October 6, 1944; October 12, 1945; December 7, 1945; April 26, 1946; May 3, 1946; October 4, 1946; November 1, 1946; November 8, 1946; November 15, 1946; February 14, 1947; February 28, 1947; March 14, 1947; May 2, 1947; May 23, 1947; September 26, 1947; October 10, 1947; October 24, 1947; November 14, 1947; November 21, 1947; April 2, 1948; April 23, 1948; October 20, 1948; April 4, 1951; May 23, 1952; March 5, 1954; November 5, 1954; April 13, 1956; April 27, 1956; May 27, 1963; September 22, 1967.

Macon (MO) Chronicle-Herald. August 25, 1944; September 19, 1947.

Missouri Herald (Hayti, MO). July 29, 1965.

Pittsburgh (PA) Courier. December 3, 1943.

Sedalia (MO) Democrat. February 6, 1977.

Springfield (MO) News-Leader. October 10, 1969.

St. Louis (MO) Post-Dispatch. May 30, 1946; December 12, 1989; November 4, 1999; February 2, 2000.

Sunday News and Tribune (Jefferson City, MO). February 6, 1966.

Chapter 13

Fulbright-Powell, Fina. Interview with author, July 1, 1923.
North Carolina Central University, yearbook, 1948. Via Ancestry.com.
United Press International. "Pope Arrives in Colombia." August 22, 1968.

Asheville (NC) Citizen-Times. August 18, 1967.
Pittsburgh (PA) Courier. April 1, 1967; August 17, 1968; August 24, 1968.
Springfield (MO) News-Leader. October 14, 2012.

INDEX

ABOUT THE AUTHOR

Michelle Brooks has made a career out of finding the lesser-known stories of the Jefferson City community and its history. For nearly twenty years, she shared those stories as a reporter for the *Jefferson City News Tribune*. Since 2021, she has compiled stories of the past into six books, this one and *Murder & Mayhem Jefferson City*, *Hidden History of Jefferson City*, *Lost Jefferson City*, *Interesting Women of the Capital City* and *Buried Jefferson City History*. Brooks' goals are to make true stories from an earlier time accessible and to celebrate the accomplishments of everyday people. She graduated from Lincoln University of Missouri in 2018 with a Bachelor of Liberal Studies degree, with emphasis in anthropology and history. Since 2019, she has worked at the Missouri State Archives in Jefferson City. When not writing or researching, Brooks enjoys gardening, music and spending time with her pets and family.